The Zone System for 35mm Photographers, Second Edition

A Basic Guide to Exposure Control

Carson Graves

Focal Press
Boston Oxford
Johannesburg Melbourne
New Delhi Singapore

Focal Press is an imprint of Butterworth-Heinemann.
A member of the Reed Elsevier group

Recognizing the importance of preserving what has been written, Butterworth-Heinemann prints its books on acid-free paper whenever possible.

Library of Congress Cataloging-in-Publication Data
Graves, Carson, 1947–
The zone system for 35mm photographers : a basic guide to exposure control / Carson Graves. -- 2nd ed.
p. cm
Includes index
ISBN 0-240-80203-9 (pbk. : alk. paper)
1. Zone system (Photography) 2. 35mm cameras. I. Title.
TR591.G73 1996
771--DC20 96–17703
CIP

British Library Cataloguing-in-Publication Data
A catalogue record for this book is available from the British Library.

The publisher offers special discounts on bulk orders of this book.
For information, please contact:
Manager of Special Sales
Butterworth-Heinemann
313 Washington Street
Newton, MA 02158-1626
Tel: 617-928-2500
Fax: 617-928-2620

For information on all Focal Press publications available, contact our World Wide Web home page at: http://www.bh.com/fp

10 9 8 7 6 5 4 3 2 1

Printed in the United States of America

Respectfully dedicated to Arnold Gassan,
teacher and friend

Contents

Preface

One of the greatest frustrations in teaching photography is watching students struggle to make the photographic materials and processes work for them. Good ideas often have trouble reaching their final form intact. The ideas either become compromised or a student becomes frustrated and simply stops using photography creatively.

Soon after I began to teach photography I realized that what my students needed was a framework upon which they could build a basic understanding of how films, developers, and printing papers functioned. Such a framework should allow them to take anything that they saw or felt and translate it into practical working methods. When I discovered the zone system, which had been devised by Ansel Adams and Fred Archer for explaining and manipulating photographic materials, I realized that it was potentially the tool that I wanted for my students.

The first time I tried to teach the zone system, however, I became so hopelessly confused in the middle of my lecture that I had to excuse the class and then sit down to figure out what I had really been trying to say. The problem was that the zone system had been presented to me by the texts of the day as a recipe for advanced photographers that unaccountably mixed higher forms of mathematics and an esoteric practice called sensitometry with the basic concepts about how film is exposed and developed. To make matters worse, most books on the subject emphasized procedures that produced a particular result and did not permit the reader to decide what was personally suitable. In short, the zone system seemed to be reserved for only those initiated few who were willing to wade through logarithms and to work with specialized equipment, and not for the average person interested in photography.

From my initial confusion, and through much trial and error, I discovered a way to teach the zone system to beginning students who only had 35mm cameras with built-in light meters. I found that not only was it possible for a beginner to master the zone system, but that it was easier to learn correct technique from the start than to unlearn incorrect technique at a later date.

The most startling discovery about the zone system, however, came only after I had been teaching it for a while. As my students used the system, it started functioning for them as a new language might, to define ideas and feelings, and relate them to the technical aspects of photography. This language of the zone system became the connection between subjective vision and the procedures necessary to produce an image. As a result, these students not only became technically better photographers but more creative as well. It is my hope, in writing this book, that this experience will now be shared by a greater number of people.

Finally, I would like to thank all of the students who have so patiently listened to my explanations of the zone system and who have asked the questions that have helped me to clarify what I was trying to say.

PREFACE TO THE SECOND EDITION

One of my most pleasurable experiences since the first edition of *The Zone System for 35mm Photographers* was published in 1982 has been the discovery that not only is the zone system timeless but clear expositions of it have lasting power, in spite of the explosive growth of automatic cameras that attempt to decide everything for the photographer from focus, to shutter speeds, to apertures. Apparently, amid all the marketing hyperbole there is still a thirst for understanding and controlling the photographic process rather than letting process live in the realm of techno-magic and thus control the photographer.

Possibly the best kept secret of the first edition is that *The Zone System for 35mm Photographers* is a book for all beginning students of the zone system, regardless of the camera and film format they use. The explanations of the zone system, how to determine exposure and contrast in a scene, and how to test films and developers are the same regardless of equipment type. Only the chapter on shortcuts for use with 35mm cameras applies specifically to roll film, and even most of these techniques are adaptable to all cameras and formats.

For the second edition, I have completely reorganized the chapters and updated the language and explanations of the various zone system tests and procedures. The basic methodology of the first edition, however — that of clearly leading readers through the theory and practice of the zone system — remains untouched.

As with all significant projects, this book contains the contributions of many people to whom I am indebted and can never adequately thank. First and foremost is the gratitude I owe my wife, Judy Canty, for years of patience and forbearance, in addition to her excellent photographic contributions. At Focal Press, Marie Lee, Tammy Harvey, Maura Kelly, and Kathryn Geiger have been instrumental in piloting this edition through the editorial and production shoals. Kathrin Unger did a superb job of editing my manuscript. Other people who have helped in many crucial ways are Margo Halverson, who provided the beautiful interior layout; Rosalyn Reiser, who made many helpful suggestions for improving the text; Cheryl Sacks, Kimberly Van Dyke-Fitch, and Callie Barnwell, who were willing and patient models for illustrations;

Tim Barnwell, who would stop everything to ship prints to me; Charles Melcher, who patiently explained color theory to me; and all the photographers who gave their images for illustrations and whose contributions are listed in the back of this book. Other help and support came from S. Randall McLamb, Karen Sadowski, and Judith Wolfman.

Finally, I would like to thank the following teachers, who along with their students supported this project with their interest and contributions of photographs for illustrations: Bill Byers at Worcester State College, Klaus Schnitzer at Montclair State University, and John Eide at the Maine College of Art.

Chapter 1

What Is the Zone System?

Photography is a medium defined by its materials and procedures. Rather than allowing this to impose limitations, photographers can work creatively with their materials. In this image, the mechanical nature of the shutter as a measuring device for time creates a gesture and form that existed only in the imagination of the photographer. In the final analysis, the only limitations are the self-imposed rules that photographers place on their creativity, not the technical nature of the medium.

When photographers first encounter the zone system, they are often overwhelmed and rightfully put off by complex descriptions of exposure calculations and the jargon of logarithmically derived film densities. This is because the zone system is frequently defined in terms of techniques rather than goals. While the technical aspects are undeniably important, why one should learn about the zone system and apply it to photography has very little to do with becoming a more accomplished technician.

Certainly, there are many technical approaches to the zone system. They vary in complexity and with the specialized needs of a particular camera or film format. There is, however, but one goal that defines the zone system, and that goal is both simple and profound: to serve as a bridge between your creativity and the techniques needed to translate that creativity onto film and paper.

The zone system bridges the gap between aesthetic choices and craft by giving you a precise language that you can use to relate your vision of a scene to the way your camera sees that scene. By clearly defining this link between you and your materials, the zone system encourages you to be more conscious about the choices you make when you photograph.

THE PLACE OF CRAFT IN PHOTOGRAPHY

Not having an accurate definition of the zone system helps explain why many photographers dismiss it as being a tool for those overly concerned with craft at the expense of self-expression. This misunderstanding of the zone system leads to a misunderstanding about the role of craft in photography, namely, that craft can serve as a substitute for vision. In fact, technical proficiency by itself does not make good images, it only makes visual failures easier to see. As some have observed, the roadside of the history of photography is full of "clear images of fuzzy concepts."

Adding to the misunderstanding about craft is the fact that producing a photographic image involves a relatively simple set of procedures. Over the years, manufacturers have taken much of the labor out of the process by putting film, developers, and papers in convenient and standardized packaging. Cameras, too, continue to have more and more automation, which makes them easy to use even if the photographer has little or no knowledge about how they work. The result is that most student photographers can make a basic print during the first week of a beginning photography class and few see the need to go beyond that ability.

Yet an understanding of process is essential to controlling the technology in one's art. The real issue is to use this craft not as an end unto itself, but as a way to understand the tools and materials of photography for the effect they have on the image. This understanding results in more choices and greater freedom for a creative photographer.

This is true in all areas of photography, not just the zone system. For example, the shutter is a mechanical device that controls the amount of light entering the camera and striking the film. Beyond that simple fact are a number of choices about how to use the shutter, each of which will produce a very different looking image. A fast shutter speed will stop moving action or render a handheld shot sharp. A slow shutter speed will cause movement to be seen as a blur, thus affecting the visual content of that movement and entirely changing its emotional appeal.

This is part of what makes photography unique as an expressive medium. With painting, for example, the amount of time it takes a painter to produce an image does not alter your experience of it, but the amount of time it takes to expose film has a profound effect on your experience of a photograph.*

* Charles Harbutt, "The Shutter," in *Modern Photography* (February 1976, p. 94).

This cityscape shows how a sensitive approach to craft can create an expressive photograph. The glowing sensation of light on the side of the building was first felt by the photographer and then translated into craft decisions about film exposure and development. The photographer's feelings about this scene could not appear successfully in the print if it were not for an understanding of the technical processes of photography.

Craft and the Zone System

In the zone system, the process of exposure and development follows exact physical and chemical laws. By understanding these laws, you can predict their effect on the final image and thus gain greater control over the photographic process.

The approach this book takes to the zone system is an empirical one. This means that you can verify any theoretical information in the text by actual experience. Simple tests are outlined to provide this experience and help you calibrate your equipment. Once you have a basic understanding, you can make craft decisions based on the zone system that are as simple or complex as you want.

As a final note, the zone system is most useful with black-and-white film. However, all photographic emulsions that are silver based, including color transparency, color negative, and even the so-called silverless or chromogenic films, function according to the principles of the zone system. A chapter about color emulsions, along with special tests for them, is part of this book. Once you understand the zone system with one film, however, you can use that knowledge with virtually all other photographic materials.

HOW TO USE THIS BOOK

The Zone System for 35mm Photographers is organized into three broad sections: theory, practice, and special applications. You should feel comfortable with the material in one section before proceeding to the next.

- Chapters 2 through 7 present the theory of the zone system, from naming the zones, to measuring light and determining contrast, to explaining how to make the zone system work for you.
- Chapters 8 and 9 outline different tests you can use to calibrate your equipment for the zone system. These tests assume that you are using a camera that allows you to manually adjust f-stops and shutter speeds.

- Chapters 10 and 11 discuss special applications of the zone system for 35mm cameras and color films.
- The appendices include a glossary of zone system terms, a test to determine if your shutter is working correctly, an explanation of how to alter the DX coding of film cassettes to get the correct exposure index with automated cameras, and a comparison of the ASA, DIN, and ISO film speed rating systems.

SUMMARY

The first step in becoming successful with the zone system is to define it correctly.

- The zone system is frequently incorrectly defined as a series of complex technical manipulations.
- The zone system is, more correctly, a bridge between expression and technique, a language that you can use to translate creative choices into technical procedures.
- Most creative photographers understand that craft is no substitute for vision, but some fail to appreciate how the choices they make are influenced by photography's technical aspects.
- Something as simple and mechanical as a shutter offers creative choices about how you want your images to look. The zone system offers similar choices.
- The zone system is an empirical application of the craft of photography. The proof of the zone system is in the results you get from it.

Chapter 2

The Zones

People see photographs on two levels. The most obvious level is the subject of the image. Also important is the way that the subject is rendered tonally. In this landscape, the dark trees in the foreground are surrounded by large areas of light snow and mist that reinforce their form and mass. One of the primary purposes of the zone system is to increase your awareness of how the tones in an image help to communicate its meaning.

The first step in learning to speak the language of the zone system is to learn its vocabulary. This vocabulary consists of the name and appearance of each of the zones. Before this chapter describes the vocabulary, however, a review of the basics of film density, the relationship of the negative to the print, and how the zone system uses these concepts is in order.

NEGATIVE/POSITIVE RELATIONSHIP

The film that you put in your camera has a unique reaction to light. As visible light strikes the mixture of silver salts, bromides, and gelatin that makes up a film's emulsion, a very real but invisible change takes place. The exposed but undeveloped film is said to contain a *latent image*. As yet science has only theories about what produces the latent image, but the effect can be demonstrated repeatedly and consistently. The latent image contains the potential for a visible image on film.

The visible film image is something we know more about. It consists of particles of silver reduced from a higher electrical state (called a *halide*) during development. These silver particles have the ability to absorb varying amounts of any light that shines through the clear film base. The more reduced silver in a given area of the negative, the more it absorbs transmitted light and the darker it looks. This darkness is called *density*. The more exposure a latent image gets, the greater the potential for density when the film is developed.

When you make an enlargement from a negative, the tones that appear on the printing paper are the result of negative densities. Whether a print tone appears dark, light, or in between is controlled by these densities.

Shadows, the areas that you see as darkest in the print, are areas of the negative that receive the least amount of exposure and have the least density. *Highlights* are the opposite; they are the lightest tones in the print, but they have the greatest density in the negative. Only the gray, *midtone* areas have approximately the same visual density in both the negative and the print.

Remember that in the negative, the dark shadow areas of an image transmit the most light (that is, they look the lightest), and that highlights absorb the most light (appear the darkest).

Photographic images consist of particles of silver that absorb light. This absorption is called *density*. The more light exposure, the greater potential for density. This is true of both negatives and prints; only the arrangement of tones is reversed. The greater the density in the negative, the lighter the corresponding tone will be in the positive print.

FROM AN INFINITE GRAY SCALE TO ZONES

On a print, where you view density by reflected light rather than transmitted light, densities appear as *tones*. Print tones vary from the maximum amount of silver that can be reduced in the emulsion (known as maximum black) to an area where all you can see is the white paper base through clear gelatin. In between these extremes lies an infinitely varied, continuous range of tones.

The concept of a continuously varied gray scale is a problem when you try to use it to describe print tones. While always realizing that you are working with continuous tone materials, it is easier if you think of the gray scale as broken up into more manageable pieces. This is what the zone system does by defining a series of tones representing a particular part of the gray scale and calling them *zones*.

The tones in a print vary continuously from white to black. Every area of the image represents a part of this continuous gray scale. The zone system organizes this infinite scale into more easily recognizable segments. You can identify each tone in this photograph as representing a particular zone.

NAMING THE ZONES

The zone system gives each zone a name and a specific attribute as it appears on a print. This allows you to describe print tones in a very precise manner, one that anyone knowing the zone system vocabulary can understand. Although it might seem a simple concept, the implication of naming the tones in your prints is an important one. As the Old Testament Hebrews knew, naming a person or a concept gives one knowledge, and through knowledge, control.

Note on Terminology

When referring to zones, this book always uses roman numerals such as zone I, zone V, and zone IX. This terminology avoids confusion with camera functions that use arabic numerals, such as f/8, or 4 as a shutter speed meaning 1/4 of a second.

Maximum Black and White Zones

In learning the names and attributes of the zones, it is easiest to start at the extremes of black and white. Maximum black, the greatest amount of density possible in a print, is zone I. Paper-base white, areas of a print where developing and fixing have left only residual amounts of silver, is zone IX. There are both physical and emotional descriptions for each of these zones.

Zone I is not only maximum black, it lacks any feeling of substance, of an object existing in space. Zone IX is a paper white highlight that also lacks substance of any sort. Both zones at the extreme ends of the zone system scale are simply flat, featureless voids. One is as dark and the other as light as a particular printing paper will allow.

Dark Shadow and Bright Highlight Zones

The next zones to consider are the ones that first show a change from the two extremes of zone I maximum black and zone IX paper-base white. These are zone II, a noticeably lighter tone on a print than maximum black, and zone VIII, the first appearance of gray in a print as it darkens from the white paper base. Neither zone shows any real detail or texture; zone II is too dark, zone VIII is too light. In fact, both might be mistaken for maximum black or white, respectively, unless you compare them to known values of zones I and IX. .

Most important, though, is the emotional content of zones II and VIII. Where zones I and IX are flat and empty in feeling, zones II and VIII give the impression of depth and volume, the feeling of something being there though not yet well defined. This is an important subtlety, because a dark and mysterious shadow in zone II becomes flat and lifeless if it appears in zone I, and a glowing highlight in zone VIII becomes blank and empty if it appears in zone IX.

The best way to distinguish between zones I and II or zones VIII and IX, on a print, is to compare the suspect zone to standard zone I or IX patches. A zone I patch is a piece of paper that has been exposed to white light and given normal development to produce the darkest possible black tone. A zone IX patch consists simply of a piece of unexposed but processed print border. To use them, place the zone I patch across a suspected zone II to see if the tone is slightly lighter than the test patch. Likewise, a zone VIII should appear darker compared with a standard zone IX patch.

Zones with Texture and Detail

Zones III and VII are the first zones to show full detail and texture in the dark and light areas of the print. Zone III is often described as the "darkest detailed shadow" and zone VII as the "lightest textured highlight." These are important zones because when you look at a print you usually have an idea of which parts of the image should be the darkest with detail or the lightest with texture.

There are usually definite objects or situations that you can use for comparative reference to zone III. For example, zone III is the tone of dark shadows cast by bushes, trees, or cars on grass, dirt, and gravel. The shadows are dark but need the texture of the grass or gravel to retain a representational appearance. A person with black hair or an animal with black fur would normally have these values as dark as possible in zone III, but with visible detail and texture. In short, any object or tone that is very dark but that you want to see with detail in the print is a zone III.

Zone VII also has some specific reference points. White painted wood on a house when it is weathered and textured, for example, is a zone VII in a print (smooth white siding would be zone VIII). Unbleached, loose-weave white fabric would be zone VII compared with bleached, tight-weave sheets, which would be zone VIII. Zone VII is any object that is light but needs to appear with detail and texture.

The Midtones

The remaining zones (zones IV, V, and VI) are best learned in relation to each other. Zone V is a specific shade of gray, one that Kodak provides as its Neutral Density Test Card. You can buy this test card at most photo stores. Kodak also binds a smaller version of the test card into both the *Kodak Color Dataguide* and the *Kodak Black and White Dataguide*.

Zone V is the middle of the zone system scale. It is the tone that marks the change in the gray scale from highlights to shadows. Every zone lighter than zone V is a highlight, and every zone darker than zone V is a shadow. Zone V itself is neither a highlight nor a shadow.

From zone V you can get reference points for zone IV and zone VI. Zone IV is the lightest of the shadow zones. Such image areas as open shade (where detail should be easily visible) or brown hair are standard zone IV values.

Zone VI is the darkest highlight zone. Concrete sidewalks and Caucasian skin are standard zone VI values. With a gray card and a few reference points as a guide, you can easily distinguish zone IV and zone VI as tones one shade darker and lighter, respectively, than zone V.

PHYSICAL EQUIVALENTS FOR ZONES

Your ability to name and describe zones enables you to translate the world you see around you into the tones that appear on your prints. It helps not only to know what each zone looks like, but to have some starting points (physical equivalents taken from typical scenes) to relate to the zones. The illustrations on pages 10 through 18 show examples of each of the zones. The caption for each illustration contains a list of typical objects that normally appear in that zone in a representational print (a print in which all the tones look the way the eye normally sees them).

The illustrations and list of zone equivalents are only starting points. You should try to add other examples to the list. Doing so will increase your understanding of the zones. You can use the method described on page 94 to test your guesses. In the next chapter, *Previsualization and Placement*, you will learn how to relate your knowledge of zones to the way your camera works.

Zone I is the darkest tone that you can produce on a print. This portrait, taken in front of a non-reflective black cloth, shows the figure framed by a zone I value. Notice the difference between the zone II and zone III parts of the skeleton costume and the flat, featureless background. Any tone not reflecting enough light to register as a tone on film appears on the print as a zone I.

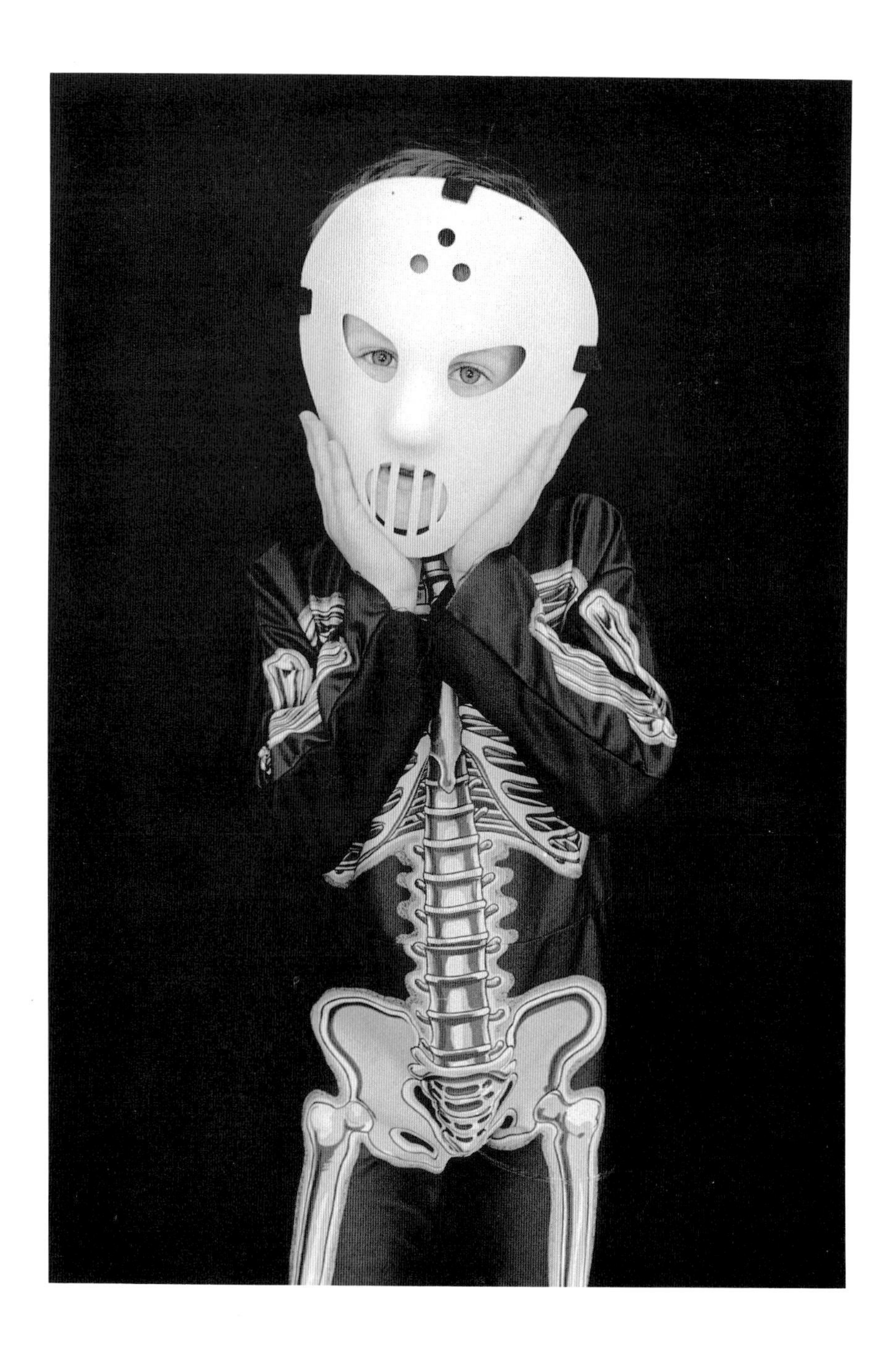

Zone II is the first tone lighter than maximum black. The water in the foreground in this image is dark and has no detail, but it has a sense of volume and substance. A zone I rendering would make it flat. Any very dark object for which you want a sense of space and volume is a zone II.

Zone III is the darkest detailed shadow in a scene. The vaulted ceiling of this interior is in zone III. You can see the architectural detail, but it retains a feeling of darkness. Zone III is any deep shadow, such as under bushes, cars, etc., in which you want to keep a sense of detail and texture. Also, black hair, black fur on animals, or black knit sweaters are typical zone III values.

Zone IV is a medium-dark tone, such as found in average dark foliage or open shade in landscapes such as the foreground of this image. Zone IV is the recommended shadow value of skin in sunlit portraits, brown hair, and unfaded blue jeans. In this example, the shadow, although dark, has the sense of being lighter and easier to see into than zone III.

Zone V is the middle of the zone system scale. Examples of zone V are a Kodak Neutral Test Card, average weathered wood, and most black skin. In this image the grass in the foreground is a zone V. The bushes appear darker because they are partially shaded.

Zone VI is the first zone lighter than middle gray. Typical zone VI values are beach sand, concrete sidewalks, and a clear north sky. Caucasian skin with light falling on it or with the subject in overall shade is zone VI. Most people recognize skin tones in a portrait that are not rendered correctly, even if they are not familiar with the zone system.

Zone VII is the lightest textured highlight in a print. In this image, the white, textured cloth hanging on the line in a zone VII. Other typical zone VII tones are blond hair, white clothes, white painted and textured wood, average snow, and cloudy bright skies.

Zone VIII is the last zone with any visible density in the print. The basketball backboard in this image, as well as any object that is smooth and very bright, is zone VIII. Like zone II, zone VIII renders objects with a sense of volume and substance. Other examples of zone VIII are smooth, painted white wood in sunlight and a piece of white typing paper.

Zone IX is the pure white of the printing paper base. Zone IX has no density and appears as a flat, featureless area in the print. The bright reflection on the polished silver is an example of zone IX, as are any specular highlights reflecting off water or other smooth surface objects.

SUMMARY

Learning the names of the zones and their attributes is equivalent to learning the vocabulary of the zone system.

- Film *density*, a product of exposure and development, is the ability of the negative to absorb light that is transmitted through it. The tones that appear on printing paper are controlled by negative densities.
- Print tones appear in one of three ways: *shadows*, the dark areas on the print produced by the least dense negative areas; *highlights*, the lightest areas of the print produced by the areas of greatest density on the negative; and *midtones*, which have approximately the same visual density in both the negative and the print.
- The zone system breaks the continuous tone gray scale into manageable segments called *zones*, gives them specific names, and defines how they look.
- Zone I and zone IX are the darkest and lightest tones, respectively, that a printing paper can produce. They are flat and featureless.
- Zone II and zone VIII are the first zones to show a change from maximum black and pure white, respectively. Although they do not show any real detail, they give the impression of depth and volume.
- Zone III and zone VII show full detail and texture in the shadows and highlights, respectively.
- Zone V is the middle of the gray scale and is equivalent to the Kodak Neutral Density Test Card. Zones lighter than zone V are highlights. Zones darker than zone V are shadows.
- Zone IV and zone VI are distinct tones that are lighter and darker, respectively, than zone V. Zone IV is the lightest shadow, and zone VI is the darkest highlight.

Chapter **3**

Previsualization and Placement

The techniques of previsualization and placement let you translate what you see into the tones that appear on the final print. In this image, it is the dark silhouette of the figures and the chairs against the lighter foreground and background that enhance the composition of the print.

If the names of the zones make up the vocabulary of the zone system, then previsualization and placement are its syntax and grammar. This chapter uses the zone definitions you learned in Chapter 2 to describe the process of previsualizing a scene and placing an exposure.

PREVISUALIZATION

Not only can you use zones to accurately describe what you see on a print, you can also use zones to describe what you see through the viewfinder when you photograph. *Previsualization* is the ability to look at a scene and translate it into the zones that you want to appear in the final print. For example, when you photograph a landscape, you might visualize a shadow under a tree as zone III, a white fence as zone VII, and the reflection of sunlight off a puddle of water in the foreground as zone IX.

In this manner, previsualization offers you a way to look at the world of colors and subtle shades, and to translate them in your mind into black-and-white print densities. In other words, previsualization is a link between you and your materials. You previsualize a scene the same way your film and paper will render it. This does not mean that you need to identify every tone in every scene that you photograph but, rather, that you label the zones that you consider most important to the expressive content of the image.

Representational and Nonrepresentational Previsualization

Imagine photographing the scene shown in the following illustration. Before you even place your eye to the viewfinder, you should look at the scene and identify the important zones in it.

This image is an example of how you can use the zone system to previsualize tones in a way that is different from the way the eye normally sees tones. The weathered wood of the dock would usually be previsualized as a zone V. Here the photographer chose to previsualize the planks as a darker tone, zone IV. This nonrepresentational previsualization created an image that draws more attention to the white cat than a representational previsualization would have.

The most important highlight is the white cat. The textured fur is an obvious choice for zone VII. The net is also a highlight, possibly a zone VI. The most important shadow tone is the wooden dock. Normally, average weathered wood is a midtone (zone V), but here the photographer has chosen to previsualize it as a darker zone to emphasize the white fur of the cat, and so has previsualized it as a zone IV.

The essential point about previsualization is that you are not limited to previsualizing a scene the way your eye sees it. A representational previsualization is only one choice you can make. If your eye sees a zone V, for example, your previsualization of it can be a zone IV or a zone VI. The versatility of the zone system allows you to previsualize zones in a way that only your imagination can see.

REVIEWING YOUR CAMERA'S EXPOSURE CONTROLS

A camera's f-stop and shutter speed controls are what you use to translate a previsualization into a photograph. Before you learn how to relate previsualization to your camera's exposure controls, however, it is a good idea to review the relationship between f-stops and shutter speeds, and how they affect an image in ways other than exposure.

F-Stops

Most photographers are aware that changing the f-stop on a lens affects the amount of light entering the lens. Fewer are aware that changing the f-stop also affects the overall sharpness of the image, something known as *depth of field*. The smaller the size of the hole (*aperture*) created by the adjustable diaphragm on the lens, the greater the apparent sharpness in both the foreground and background of the image. Beyond that, there are some basic facts about f-stops that are important to understand if you are to effectively work with the zone system.

The number of an f-stop expresses the relationship of the focal length of the lens (hence "focal" or "f" stop) to the size of the aperture. The f-number is the denominator of a fraction that states the actual size of the aperture relative to the focal length. A 100mm focal-length lens at f/2 has an aperture diameter of one-half the focal length, or 50mm.

Thinking of f-stops in this way helps clear up two points of confusion common among photographers. First, as f-stops get larger in numerical value (2, 2.8, 4, etc.), the actual size of the aperture gets smaller, since, for example, 1/2 is larger than 1/4. Second, the same f-stops on different focal length lenses have different diameter apertures.

For example, a 50mm lens at f/2 has an aperture diameter of only 25mm compared with twice that on a 100mm lens at f/2. In spite of the different size openings, both the 100mm lens and the 50mm lens gives the same exposure to a piece of film at f/2. This is true for all lenses of any focal length.

A change in an f-stop changes the exposure by a factor of two. Stopping the aperture down from f/5.6 to f/8, for example, cuts the exposure in half, while opening the aperture up from f/5.6 to f/4 doubles it. This is why the standard progression of f-numbers does not follow an arithmetic order. The fractions expressed by f-stops are calculated to give the size aperture needed to change the exposure by this factor of two.

The best way to deal with the potential confusion that f-stops cause in calculating exposure is to simply memorize the standard sequence. The sequence goes f/1.2 (the largest aperture commonly found), f/1.4, f/2, f/2.8, f/4, f/5.6, f/8, f/11, f/16, f/22, f/32, and f/45 (typically the smallest aperture on modern lenses.)

Committing this sequence to memory lets you to quickly make changes in exposure without becoming confused about how many f-stops you have changed or whether you are increasing or decreasing the exposure. It also prevents you from accidentally using an f-stop that does not change the exposure by a factor of two. For reasons best known to camera manufacturers, many lenses have a nonstandard f-stop (such as f/1.8, f/1.9, or f/3.5) as their maximum aperture. You should avoid nonstandard f-stops because they do not change exposure by an easily predictable factor such as two.

Shutter Speeds

Like f-stops, most shutter speeds are expressed as the denominator of the fraction that is the actual speed. For example, 2 is $^1/_2$ of a second, 60 is $^1/_{60}$ of a second, and so on. From one number to the next, the exposure time changes by a factor of two. The shutter speed marked 250 is twice as fast as 125, and 30 is twice the exposure time of 60.

Shutter speeds, like f-stops, also affect the image in more ways than just exposure. The description on page 2 of how choosing a fast shutter speed or a slow shutter speed can change the visual content of an image is an example of the importance of considering factors other than exposure when you choose a shutter speed.

LAW OF RECIPROCITY

Since f-stops and shutter speeds both affect exposure by a factor of two (even though they differ in the way they affect the look of the image) they are said to have a one-to-one, or reciprocal, relationship. This relationship, known as the *law of reciprocity*, means that in a given lighting situation you can keep the same exposure when you change either the shutter speed or the f-stop by making a corresponding opposite change in the other. Thus, using the law of reciprocity, you can choose to affect motion or depth of field and still maintain the same exposure for any image you make.

These images were shot at the same exposure, but using different combinations of f-stops and shutter speeds. The left-hand photograph was shot with a fast shutter speed to stop the action. A large aperture was used, and the image has very little depth of field. The photograph on the right was shot with a small aperture, and the depth of field has increased to make the background sharp. The moving figure is blurred, however, because the small f-stop required a slow shutter speed.

Reciprocity Failure

Whenever the law of reciprocity is mentioned, some photographers confuse it with the term *reciprocity failure*. What reciprocity failure in fact means is that in certain extreme situations, when the shutter speed is either very fast (above $^1/_{10,000}$ of a second) or very slow (slower than $^1/_2$ second for many films), or when light levels are very low, the one-to-one relationship between shutter speeds and f-stops breaks down. In other words, the law of reciprocity fails. In this case you must use a greater-than-predicted exposure to obtain a particular film density. In actual practice, 35mm camera users rarely encounter reciprocity failure.

PLACEMENT

Placement is how you use a camera's controls to record your previsualization on film and ultimately on the print. When you change exposure by a factor of two (one f-stop or shutter speed), you change a film's density by one zone. If there is a zone V in a scene, for example, and you want it to appear as a zone VI on your print, you simply make a change in exposure of one stop. Because zone VI is one zone lighter than zone V, you need the equivalent of one stop more exposure to increase the film's density.

This sort of manipulation is an example of how you can choose a tone in the original scene and place it in the zone that you want for the final print. The important thing to remember is that any change in exposure changes the zones that appear in the print, and any change in previsualization requires a change in exposure.

Film Density and the Number of Zones

An implied question that the relationship between zones, film density, and exposure answers is why the zone system defines nine zones instead of eight or ten. The reason is that it takes the equivalent of nine f-stops to change a film's density from what appears as maximum black (zone I) to what appears as paper-base white (zone IX) on normal contrast printing paper. Film can contain more than nine zones of density, but printing paper cannot show them. Either the brightest highlight densities will all appear in zone IX, or the darkest shadow densities will appear in zone I, or sometimes even both.

How you place zones and the limitations that the materials put on this manipulation are the subjects of the next chapter, *Measuring Light*. The important thing to remember is the relationship between zones and the camera's exposure controls. Any change you make in exposure changes the zones that appear in the final print.

SUMMARY

Previsualization and placement are the syntax and grammar of the zone system.

- *Previsualization* is the ability to look at a scene and see the zones you want to appear on the print you make of that scene.
- You are not limited to previsualizing a scene the way your eye sees it. A representational previsualization is only one choice you can make.
- *Placement* is the ability to use your previsualization and knowledge of your camera's exposure controls to affect the densities that appear on your film. This, in turn, affects the tones that occur in your prints.
- F-stops and shutter speeds both change the exposure of film by a factor of two. This relationship is known as the *law of reciprocity.*
- You can change film density by the equivalent of one zone by changing the exposure by one f-stop or one shutter speed. This is called *placing a zone*.
- The zone system defines nine zones instead of eight or ten because it takes the equivalent of nine f-stops to change a film's density from what appears as maximum black (zone I) to what appears as paper-base white (zone IX) on normal contrast printing paper.
- Film can have more than nine zones of density, but these densities will not appear as separate tones on printing paper.

Chapter 4

Measuring Light

Measuring the light reflectance in a complex scene is the first step in controlling the tones of that scene. In this image, the material of the white dress against the light skin tone required a special placement to be rendered as a zone VII. Allowing the light meter to "average" this scene would produce an underexposure and render the highlight zones much darker than previsualized.

Light is the substance of photographs. Photographers do not take pictures of objects, they actually record the light that reflects off the objects. Measuring light reflectance and understanding what to do with those measurements is the next step in learning the zone system.

WHAT LIGHT IS (AND IS NOT)

In 1803, an Englishman named Thomas Young proved, through an elegantly simple and easily repeatable experiment, that light is made up of continuous waves, much like ocean waves or the way a string vibrates when it is stretched between two points and shaken.

In 1905, Albert Einstein developed the theory that light is composed of tiny discrete particles, which he called *photons*. This theory is one of the building blocks of a study of physics called *quantum mechanics*, and it won Einstein a Nobel Prize in 1921.

Neither theory, however, can disprove the other. Whether light is made up of waves like ripples on a pond or tiny particles shot out like bullets from a gun is a paradox that physicists cannot yet resolve. One theory has even gone so far as to suggest (with appropriate mathematical proof) that light is a living organism that can make choices and respond to situations.*

REFLECTED LIGHT

Fortunately, photographers do not have to know what light is made of in order to measure it. The important thing is that light reflects off everything we see, colors as well as shades of gray. How light or dark an object appears to us depends on the amount of light that the surface of the object reflects. A white sheet of typing paper, for example, reflects about 90% of the light falling on it, whereas a piece of black velvet reflects only 5% to 6% of that same light.

Colors, too, are a function of reflected light. Whereas gray tones reflect the entire light spectrum evenly, colors reflect only a part of the visible light spectrum. Red objects, for example, reflect primarily the longer wavelengths of visible light, while blue objects reflect the shorter wavelengths. A dark color reflects a small percentage of that wavelength, and a light color reflects a greater percentage.

Every tone has a specific reflectance that determines how it looks relative to the other tones in a scene. The piece of white paper (zone VIII) reflects about 90% of the light striking it, whereas the dark cloth (zone II) reflects only about 5% to 6% of the same light and is seen as a zone II. For comparison, a gray card (zone V) reflecting exactly 18% of the light is shown between the two. The Caucasian skin (zone VI) reflects about 30% to 40% of the light.

LIGHT METERS

Photographers have two tools with which to measure reflected light. The most sensitive and easy-to-use tool is the photographer's own eye. Unfortunately, our eyes are never constant; they are always adjusting for variations in light intensity. This makes it difficult to compare the light reflected off of two different objects.

The photographer's second tool is the light meter. Most handheld and all in-camera light meters measure reflected light. These light meters have the ability to distinguish between objects that reflect different amounts of the light falling on them.

Some light meters, instead of reading reflected light, read the amount of light falling on a scene. Called *incident meters*, nearly all are hand-held devices distinguished by a white plastic dome or cone covering the light-sensitive cell. Many reflected light meters, in fact,

* Gary Zukav, *The Dancing Wu Li Masters: An Overview of the New Physics* (New York: Bantam Books, 1979).

have attachments that convert them to incident meters. The primary problem with incident meters is that they have no way to measure the relative reflectance differences between objects.

If you have, for example, zone IV brown hair and zone VII blond hair in a scene, a reflected light meter can indicate a difference between the two tones, whereas an incident light meter cannot. For this reason, incident light meters are difficult, if not impossible, to use with the zone system.

Medium Gray

Because a light meter is an electrical/mechanical device, it does not know if the light it is measuring is reflected from a shadow, a highlight, or a midtone. Instead, a meter is calibrated to compare the light it measures against a constant value. For all but some special-purpose light meters, this standard is a gray tone that reflects 18% of the light that falls on it. This means that every time a light meter gives an exposure reading, that exposure will produce an average density on film equivalent to a gray tone with a reflectance value of 18%. This is true even if what the meter measures is a detailed shadow you have previsualized as zone III, or white snow that you have previsualized as zone VII or zone VIII.

Average Meter Readings

Manufacturers of light meters know that many photographic scenes have about the same amount of dark tones as light tones in them. A typical landscape, for example, consists of approximately equal areas of sky (a bright highlight), sunlit grass and trees (midtones), and shaded areas (dark shadows). This averages to about 18% reflectance. Using a reflected light meter to measure average overall light reflectance for such scenes usually provides an adequate exposure.

The problem with exposures based on average overall meter readings is the word *average*. Average readings do not indicate an accurate exposure except in situations where the tones in a scene reflect a combined 18% of the ambient light. The consequences of relying on average meter readings might be the following: In a scene in which you previsualize mostly dark tones, an exposure based on an average reading will make that scene appear gray and take the life away from any dark, detailed zone III shadows. The same holds true for a scene composed primarily of light tones. The delicacy of a correct zone VII textured highlight is destroyed if it appears too dark and gray on the print.

If you rely only on average meter readings in situations such as these, you ultimately either accept poor exposure (and print quality) for these scenes or you learn to avoid them altogether. This can make the word *average* as much a judgment of the photographer as it is a measurement of light. As you begin to push your vision into picture-taking situations where the tones do not have an overall 18% reflectance, you must no longer accept *average* as either a lighting condition or a description of your photographs.

Indicated Meter Readings

The solution to the problem of average meter readings lies in the way the zone system relates to the light meter's 18% reflectance calibration standard. An 18% reflectance tone is the same gray tone that the zone system defines as zone V. When you use a light meter to measure light reflected from a specific, previsualized object, the reading you get is called an *indicated meter reading*. In the zone system, this reading does not directly give you the exposure you will use to make the image. If you did use the indicated meter reading as the exposure, it would produce a film density that would make that tone appear as a zone V middle gray, regardless of how much lighter or darker you previsualized it. Instead, you start with an indicated meter reading and then modify it through placement to make the final image a reflection of your original previsualization.

PLACEMENT AND INDICATED METER READINGS

Placement is what allows you to relate indicated meter readings to your previsualization. Since the relationship between each zone is an exposure change of one f-stop or shutter speed, knowing that an indicated meter reading is always zone V means that you can place that reading in any zone you want. All you have to do is determine the difference in zones between the previsualized zone and zone V. Then change the indicated meter reading by that number of stops.

For example, to get the correct exposure for a previsualized zone IV head of dark brown hair, change an indicated meter reading of 1/60 second at f/8 to 1/60 second at f/11 (or 1/125 second at f/8 depending on whether you want greater depth of field or stopped action). If you did not place the indicated meter reading, the dark brown hair you previsualized in zone IV would end up as a zone V film density. Stopping down one f-stop or shutter speed produces a zone IV film density. Indicated meter readings and placement are the tools you use to relate the way your eye sees reflected light (previsualization) to the way the light meter and film see reflected light (exposure).

Examples of Placement

Pages 30 and 31 contain examples of a specific tone, first unplaced and left in zone V and then placed into its previsualized zone. In each case, the placed exposure was changed from the indicated meter reading by counting the number of zones that the previsualized tone was different from zone V, and then changing the indicated meter reading by that number of stops. More exposure was used if the previsualized zone was a shadow, and less exposure was used if the previsualized zone was a highlight.

Shadow Left in Zone V

Shadow Placed in Zone III

The shadow in the two images of the tree was previsualized in zone III and a meter reading was made of that tone. The first illustration shows the shadow left in zone V, and the second illustration shows the shadow placed in zone III. To place a zone III shadow, take the indicated meter reading from that tone and stop down two stops.

Face Left in Zone V

Face Placed in Zone VI

The face in the two portraits was previsualized in zone VI, and a meter reading was made of that tone. The first illustration shows the face left in zone V, and the second illustration shows the face placed in zone VI. To place a zone VI highlight, take the indicated meter reading from that tone and open up one stop.

Snow Left in Zone V

Snow Placed in Zone VII

The snow in the two landscapes was previsualized in zone VII and a meter reading was made of that tone. The first illustration shows the snow left in zone V, and the second illustration shows the snow placed in zone VII. To place a zone VII highlight, take the indicated meter reading from that tone and open up two stops.

USING A CAMERA'S BUILT-IN LIGHT METER

Successfully placing an exposure requires knowing how to make accurate light meter readings. Many people believe that the zone system requires the use of a special type of hand-held light meter called a *spot meter*. Spot meters make it easier to select a specific part of a scene to measure. While it is true that in some situations hand-held light meters offer certain conveniences, your camera's built-in light meter is just as capable of accurate readings if you use it correctly.

The light meter built into most 35mm SLR cameras, like most light meters, is calibrated to average the tones found in the viewfinder to an 18% reflectance value. This includes "center weighted" metering systems, which only limit the area in the viewfinder that most of the reading is made from, and the "programmed" metering systems, which go one step further by deciding for you which part of a scene is more important to meter than another.

You can use the following procedure with your camera's built-in light meter to get accurate meter readings from a previsualized tone:

1. Isolate a previsualized tone so that the meter reads only the light reflected from that area. This means getting close enough to that part of the scene so that it completely fills the viewfinder. The object does not have to be in focus.
2. Adjust your camera's f-stops and shutter speeds to match those indicated by the built-in light meter. This is the indicated meter reading.

3. Place the indicated meter reading into your previsualized zone.
4. Step back to frame the entire scene and, ignoring any further changes in what the light meter indicates, take the picture.

Be aware that the area of the previsualized zone must be large enough to fill your viewfinder without forcing the camera so close that the lens (or the photographer) casts a shadow and causes a false reading. With a normal (50mm) lens, a good minimum size for a previsualized area is 8 × 10 inches. If you are using a wide-angle lens and it prevents you from isolating a single tone, then substitute a longer focal length lens to take the reading. You can switch back to the wide-angle lens for the exposure.

Light meters built in to 35mm cameras read the light reflected from all the objects that appear in the viewfinder. To accurately read the light reflected from a single tone, you must get close enough so that the tone completely fills the viewfinder. In this illustration the photographer is taking an indicated meter reading from the shadow on the side of the rock.

SUMMARY

Indicated meter readings and placement are the tools you use to relate the way you previsualize a scene to the way film records light.

- Film records light reflected off the objects in a scene. How light or dark these objects appear on the film is a function of the percentage of light that each object in the scene reflects.
- Reflected light meters average everything they measure into an 18% reflectance gray tone, which is a medium gray equal to zone V.
- An indicated meter reading is a reflected light reading from a previsualized zone. An indicated meter reading renders the tone as a zone V medium gray.
- To get the correct exposure for a scene, take an indicated meter reading from a previsualized tone and place it by changing your camera's settings one stop away from the indicated meter reading for each zone that the previsualized tone differs from zone V.
- Use your camera's built-in light meter to get accurate indicated meter readings by walking up to the subject and isolating a previsualized tone in the viewfinder before you take the reading.

Chapter **5**

Light and Contrast

The only difference in the photographs illustrated here is the intensity of the light. Being sensitive to the quality of light in a scene is key to using the zone system in a conscious, creative manner.

Light is constantly changing, and as it changes, it has the ability to alter a scene even when everything else remains the same. As the light in a scene changes, so does the difference between the amount of light reflected from highlights and the amount of light reflected from shadows, a difference known as *contrast*. To account for these changes, the zone system includes a language that precisely defines the relationship between previsualization, reflected light, and contrast.

PREVISUALIZATION AND LIGHT REFLECTANCE

Previsualization is the process of deciding how you want the tones in a scene to appear on the final print. The amount of light reflected from those tones, or their *light reflectance*, determines how they will actually appear. Your previsualization and the light reflectance will not always match. Although you may visualize two tones as being four zones apart (such as a zone III and a zone VII), the zone VII tone will not always reflect four stops more light than the tone in zone III. It is the intensity of the light falling on the scene (the *ambient light*), rather than previsualization, that affects what the meter reads.

CONTRAST AND LIGHT INTENSITY

To understand this, imagine an arbitrary unit of light. Suppose 10 of these light units are falling on a scene in which you previsualize a zone III (darkest detailed) shadow and a zone VIII (bright with little or no texture) highlight. The zone III reflects approximately 10% of the light falling on it, while the zone VIII reflects approximately 90%.

When there are 10 units of ambient light, the zone III reflects 10% of that, or 1 unit of light. The zone VIII reflects 90%, or 9 units of light. Since contrast is the difference between the highlight reflectance and the shadow reflectance, you could say that the contrast for this scene is 8 units of light.

Now, increase the amount of ambient light in the scene to 100 units. Your previsualization of the scene hasn't changed; you still want the two tones in the final print to be zone III and zone VIII. The percentage of light that the objects reflect also has not changed. The only change is in the intensity of the light. But this change means that the zone III now reflects 10 units of light and the zone VIII now reflects 90 units.

This makes the difference between the two previsualized tones 80 units, a 10-fold increase over the first example. So, instead of contrast being a constant, it is actually relative to the overall intensity of light. You should always expect a different contrast on a bright, sunny day than at dusk or when the light intensity is low.

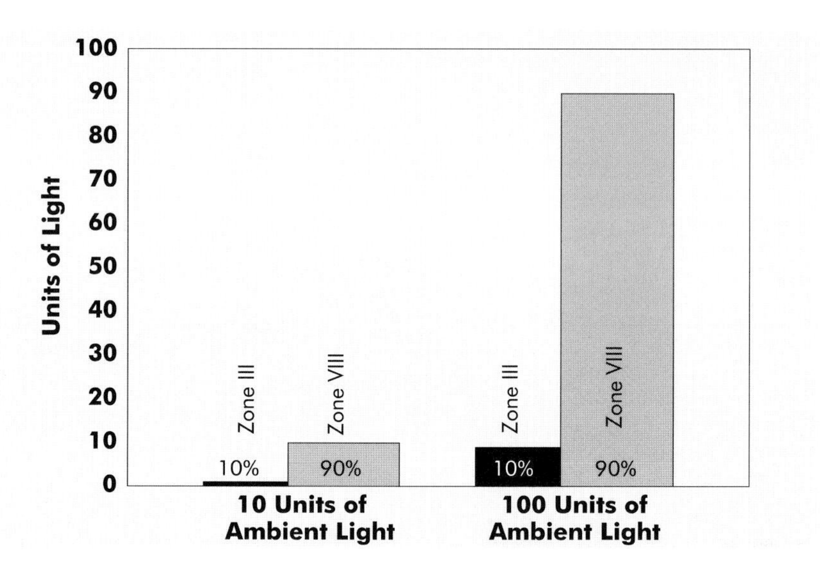

Contrast Range In a situation where one tone reflects 10% of the light falling on it and another tone reflects 90%, the intensity of ambient light determines the contrast. This graph shows that when the ambient light is 10 units, the difference in the light reflectance is 8 units. When the ambient light is increased to 100 units, the difference in reflected light between the same two tones increases to 80 units.

TYPES OF CONTRAST

Three terms describe the types of contrast that can occur in a scene:

- *Normal contrast*, which occurs when the difference between indicated meter readings of highlights and shadows is the same as the difference in the previsualized zones.
- *Low contrast*, which occurs when the difference between indicated meter readings of highlights and shadows is less than the difference in the previsualized zones.
- *High contrast*, which occurs when the difference between indicated meter readings of highlights and shadows is greater than the difference in the previsualized zones.

The following sections describe the subjective appearance of normal, low, and high contrast scenes and provide a way for you to measure their contrast.

Normal Contrast

Normal contrast in a scene has a visual appearance of an even range of tones, not very dark in the shadows and not too brilliant in the highlights. Shadows cast by directional light appear fully defined but not harsh. There is an overall pleasing, comfortable feeling from the light.

You can usually find a normal contrast scene in the open shade of the north side of a building on a sunny day, or out in the open on an overcast day between the hours of 10 A.M. and 3 P.M. Remember, however, that when you are evaluating contrast, the way a scene appears to your eyes is subjective. Although experience and observation are useful in getting a general idea of the contrast of a scene, the only true indication is with careful light meter measurements.

Normal Contrast occurs when the number of stops between indicated meter readings of the most important highlight and shadow tones equals the difference between the number of previsualized zones. In this image, the zone IV previsualized for the brown hair had an indicated meter reading of f/5.6 at 1/125, and the zone VI skin tone had an indicated meter reading of f/11 at 1/250. These indicated meter readings are two stops apart, and the previsualization is also two zones apart. Hence, the scene has normal contrast (N).

Testing for Normal Contrast

Determine the contrast of a scene by comparing indicated meter readings from at least one previsualized highlight and one previsualized shadow. In a normal contrast scene, the difference between meter readings of a highlight and shadow should correspond to the difference between the zones.

For example, when you previsualize two tones as zones IV and VI, normal contrast is indicated when your meter reads exactly two stops of exposure difference between the light reflected from the two tones. Likewise, when you previsualize two tones as zones III and VII, normal contrast is indicated when your light meter reads exactly four stops of exposure difference. When testing for contrast, always compare a shadow (zones I through IV) to a highlight (zones VI through IX). The differences within shadows or within highlights only are not enough to accurately determine the contrast of a scene. In the zone system, the shorthand notation that is used to describe normal contrast is the letter "N."

The most convenient way to express the difference between indicated highlight and shadow meter readings is in f-stops, or just "stops," even though most light meters indicate exposure with both a recommended f-stop and a shutter speed.

Low Contrast

Low contrast in a scene occurs when the reflectance values are less than one stop for each previsualized zone. To the eye, a low contrast scene is one of even, nondirectional light. Shadows cast by objects are either soft and indistinct or nonexistent. Low levels of light intensity most often produce a low contrast scene, but not always.

A notable exception is a sunny landscape covered by snow or beach sand. Because snow and light sand are highly reflective, a sunny day will "bounce" light into the shadows, filling them in. The result is frequently a low contrast scene.

Testing for Low Contrast

As with other types of contrast, the only way to make more than an educated guess about low contrast is to measure a scene with a light meter. Meter readings from previsualized highlights and shadows in a low contrast scene indicate less of a difference in stops than zones. In a scene with a previsualized zone III and zone VIII, for example, you would find only a four stop or less difference in the indicated meter readings of these two tones.

The shorthand notation used to describe low contrast is "N+," to indicate that there are more previsualized zones than stops. If there is one extra zone, such as a scene in which there are only three stops difference between previsualized zones III and VII, the contrast is N+1. When there are two extra zones, such as a three stop difference between previsualized zones III and VIII, the contrast is N+2, and so on.

Low Contrast occurs in a scene where the number of stops between the indicated meter readings of the most important highlights and shadows is less than the number of previsualized zones. In this image, the dark centers of the leaves were previsualized as zone IV and had an indicated meter reading of f/5.6 at $^1/_{125}$. The white edges of the leaves were previsualized as zone VII and had an indicated meter reading of f/11 at $^1/_{125}$. The indicated meter readings are two stops apart, while the previsualization is three zones apart. Hence, the scene has low contrast (N+1). This illustration is printed to show the low contrast; on page 85 is an example of the image printed as it was previsualized.

High Contrast

High contrast in a scene occurs when the reflectance values are greater than one stop for each previsualized zone. Such a scene appears to have strong intense light in that it casts dark, sharply defined shadows. Most bright sunlit scenes are high contrast (except for the previously mentioned landscape scenes of snow or sandy beaches); so, too, are scenes lit by bright artificial lights, such as stage productions. Any scene in which the range of tones goes from very bright in the highlights to very dark in the shadows is usually a high contrast scene.

Testing for High Contrast

When you compare indicated meter readings in a high contrast scene, the difference in stops is greater than the difference in previsualized zones. For example, a scene with previsualized zones III and VIII will measure six or more stops difference, and a scene with previsualized zones IV and VII will measure at least five stops difference.

The shorthand notation used to describe low contrast is "N−," to indicate that there are fewer zones than stops. If there is one less zone, for example, a scene in which you measure five stops between previsualized zones III and VII, the contrast is N−1. When there are two fewer zones, such a scene in which you measure seven stops between previsualized zones III and VIII, the contrast is N−2.

High Contrast occurs in a scene when the number of stops between indicated meter readings of the most important highlights and shadows is greater than the number of previsualized zones. In this image the facade of the house was previsualized in zone IV and had an indicated meter reading of f/8 at 1/125. The sunlit road was previsualized in zone VIII and had an indicated meter reading of f/22 at 1/500. With five stops between the two tones but only four zones between their previsualizations, the scene is high contrast (N−1). This illustration is printed to show the high contrast. On page 86 is an example of the image printed as it was previsualized.

PREVISUALIZATION AND CONTRAST

Changes in previsualization affect contrast just as much as changes in light intensity. In fact, it is possible to have more than one previsualization for a single scene. The illustrations on page 39 show how a simple change in previsualization changes a low contrast scene to a high contrast scene.

Even when ambient light remains the same, a different previsualization can mean a different contrast. In the top image, the shadows in the vines were previsualized as zone IV and the highlights on the vines as zone VII. The indicated meter readings of these tones were only two stops apart, so that the scene was N+1 (low contrast). In the bottom image the black cloth was previsualized as zone III and the shirt as VIII. Here the indicated meter readings of these tones were five stops apart. This made the scene N (normal contrast). What changed was the previsualization, not the light intensity and not the reflectance of the vines and the background.

Practicing Previsualization

As you learn the zone system, you will constantly adjust how you previsualize scenes. You may find yourself making prints in which the tones appear as you previsualized them, but the image still does not possess your feelings for the scene at the time you exposed the film. This is because the translation of your mental equivalent for each zone into actual print tones is a process that you can fully understand only through practice.

When failure does occur, ask yourself whether you really previsualized the shadows and highlights the way you wanted them. Perhaps, as an example, you might have gotten a more expressive print if you had previsualized a highlight as zone VII instead of zone VIII, utilizing the detail of the highlight area instead of just rendering an object with no definite texture. Changing the previsualization of the highlight in this case changes the contrast of the scene. Or, perhaps you might have previsualized a shadow in zone III instead of zone IV, giving the maximum effect of a dark shadow while still retaining detail. Changing your previsualization of the shadow in this case changes both the contrast of the scene and your exposure placement.

As you become comfortable with the zone system, keep examining your prints and try to visualize how the tones in them might have appeared had you previsualized the scene in a different way. Always allow yourself the freedom to explore different possibilities. In Chapter 7, *Making the Zone System Work*, you will learn ways to try out different previsualizations while photographing a scene, which will help the learning process go faster.

SUMMARY

The zone system includes a method for describing contrast that enables you to relate indicated meter readings to your previsualization of highlights and shadows.

- Light intensity affects contrast by changing the total amount of light reflected from the objects in a scene. Generally, the greater the light intensity, the greater the contrast.
- While certain qualities of light are typical of a particular type of contrast, the only way to be sure of the contrast of a scene is to compare indicated meter readings to your previsualization of highlights and shadows.
- Normal contrast describes a scene in which the difference between indicated meter readings is the same as the difference between previsualized zones. Use the notation, "N," as a shorthand for normal contrast.
- Low contrast describes a scene in which the difference in indicated meter readings is less than the the difference in previsualized zones. The notation for low contrast is N+.
- High contrast describes a scene in which the difference in indicated meter readings is greater than the difference in previsualized zones. The notation for low contrast is N−.
- Previsualization affects contrast as much as light intensity affects contrast. You can alter the contrast of a scene just by changing your previsualization, even if the light intensity and the reflectance values of all the objects in the scene remain constant.
- Translating the concepts of previsualization and contrast into an actual print is a process you can only fully understand through practice.

Chapter 6

Development and Contrast

The back lighting (and the pre-visualization), created a scene with high contrast. Without losing the sense of light, the negative densities were made printable by modifying the development of the film.

UNDERSTANDING EXPOSURE AND DENSITY

In the nineteenth century, photographers had to make their own film emulsions. Daguerreotypes, wet plates, even the first dry plates were all to a certain extent fabricated by the individual photographer, often just before exposing. There was no standardization for these processes; each photographer had a jealously guarded formula worked out through trial and error. Exposure was intuitive and most photographers felt that through development they could alter the negative at will, correcting for mistakes in exposure and changing the relationships of tones any way they wished. Because there was no standard to observe, no one noticed how exposure and development actually affected the final image.

Toward the end of the nineteenth century, when consistent film emulsions were first manufactured, attitudes about film development were forced to change. Two amateur photographers and industrial chemists became interested in the new film emulsions.

These scientists, named Ferdinand Hurter and Vero Charles Driffield, turned an old sewing machine and the light from a single candle into an apparatus that demonstrated for the first time the effect of exposure and development on film density.*

What Hurter and Driffield discovered shocked the photographic world of their time. First, they proved that a given film emulsion had a specific sensitivity to light that development could not alter. Second, they found that once a negative was correctly exposed, there was but a single development that created the optimum separation of tones for a particular scene. In other words, a negative's density had a specific and predictable relationship to exposure and development.

Exposure and Shadow Density

Exposure, as Hurter and Driffield found, controlled the density of the shadows, the areas of the negative that receive the least exposure. Contrary to what photographers believed at the time, no amount of additional development could add density to an underexposed shadow.

Changes in development, whether in time, temperature, chemical concentration, or even patterns of agitation, did cause changes in areas of the negative that received the most exposure. These were the midtones and highlights. These tones changed density in proportion to the amount of exposure. The greater the exposure, the greater the potential for change during development.

Negative Densities After Development

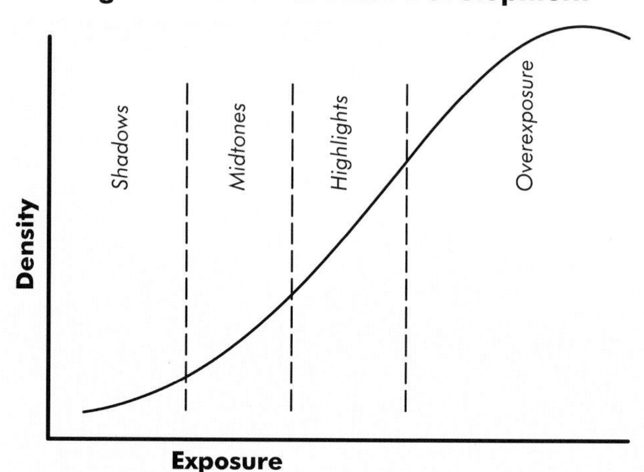

This graph, based on the research of Hurter and Driffield, shows that the lowest densities (shadows) do not respond significantly to development. The part of the curve representing those tones is relatively flat. Midtones and highlights, on the other hand, are increasingly affected by development, as shown by the progressively steeper curve. Interestingly, as the graph indicates, Hurter and Driffield found that as exposure increases far beyond normal, density stops increasing and actually begins to decrease.

The graph shows this effect. The horizontal axis, reading from left to right, represents increasing exposure, and the vertical axis shows increased film density with increasing height. Where the two meet is zero. The curve, known today as a Hurter and Driffield (or H&D) curve, represents a typical negative's densities. It has a shallow, gradually increasing slope in the areas of least exposure (the

---

* W.B. Ferguson, ed., *The Photographic Researches of Ferdinand Hurter and Vero C. Driffield* (Dobbs Ferry, NY: Morgan & Morgan., 1974) and Beaumont Newhall, *The History of Photography* (New York: The Museum of Modern Art, 1982).

shadows), and a steeper, though relatively proportional, slope as exposure increases in the midtones and highlights.

Hurter and Driffield's findings are expressed in the old zone system adage "expose for the shadows and develop for the highlights." This simple statement is an acknowledgment that you control shadow densities through exposure placement and you control highlight densities by changing film development.

HOW DEVELOPMENT AFFECTS SHADOWS

As Hurter and Driffield's research indicates, exposure placement controls the appearance of shadows in a negative. Development does not have much effect on these densities. Looking at idealized density curves on a graph and comparing these curves to actual negatives illustrates why.

The following graph shows the effect of development on shadow densities. The horizontal axis represents increasing development, and the vertical axis represents increasing film density. A curve representing the density of a typical shadow exposure is essentially a flat line. As development increases along the horizontal axis, the shadow density does not move very far up the vertical axis. As the dotted lines in the graph indicate, the only way to significantly increase or decrease the density of a shadow is to change exposure, in effect placing the shadow in a different zone. The dotted lines represent two different possible placements, one higher and one lower. Even with a different placement, changes in development do not significantly change a shadow density.

Effect of Development on Shadow Density

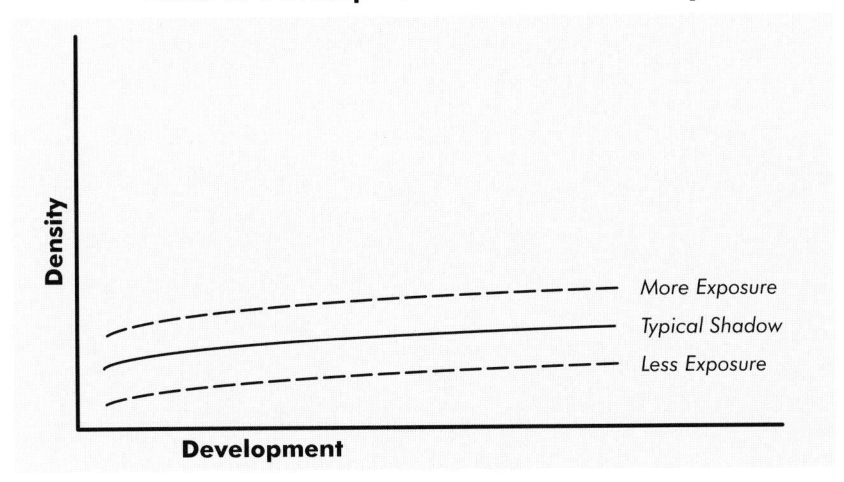

This graph illustrates the effect of changing development on shadow densities. The curve representing a typical shadow density is essentially a flat line, increasing only slightly even as development increases. The only significant increase or decrease in shadow density comes from changing exposure placement.

Underexposed Negative

Correctly Exposed Negative

Overexposed Negative

The three negatives above illustrate the graph of the shadow densities. Look at the black drape over the model's shoulders, which was previsualized as a zone III. The underexposed negative shows little or no detail in the zone III drape, meaning that it will look more like a zone II on the print. The correctly exposed negative renders the zone III with adequate detail. The overexposed negative shows plenty of detail but also has extra density, which will appear as extra grain in the print.

HOW DEVELOPMENT AFFECTS HIGHLIGHTS

A graph showing what happens to highlight densities during development looks much different than the graph of shadow densities. The curve goes up at a steep angle. Starting at about the same density as the shadows, a highlight rapidly builds up density as development increases. What you see in the following graph is that without changing exposure, you can change the density of a highlight by changing its development.

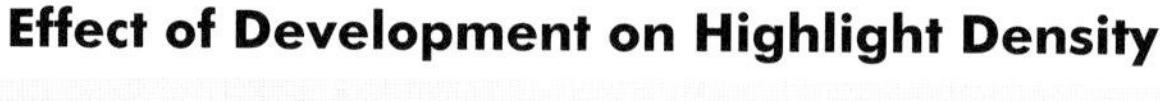

Effect of Development on Highlight Density

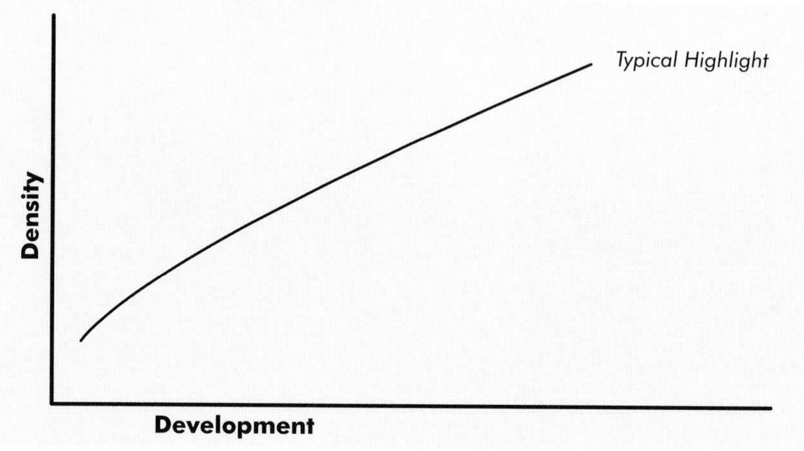

This graph illustrates the effect of changing development on highlight densities. The curve representing a typical highlight has a steep slope, building up density as development increases.

HOW DEVELOPMENT AFFECTS CONTRAST

When you combine the graph of the shadow densities with the graph of the highlight densities, the result shows how development affects a negative's contrast. As development increases, so does the difference between the shadow densities (which remain relatively constant) and the highlight densities (which increase at a steady rate). In other words, the longer you develop film, the higher the contrast of the negative.

Effect of Development on Contrast

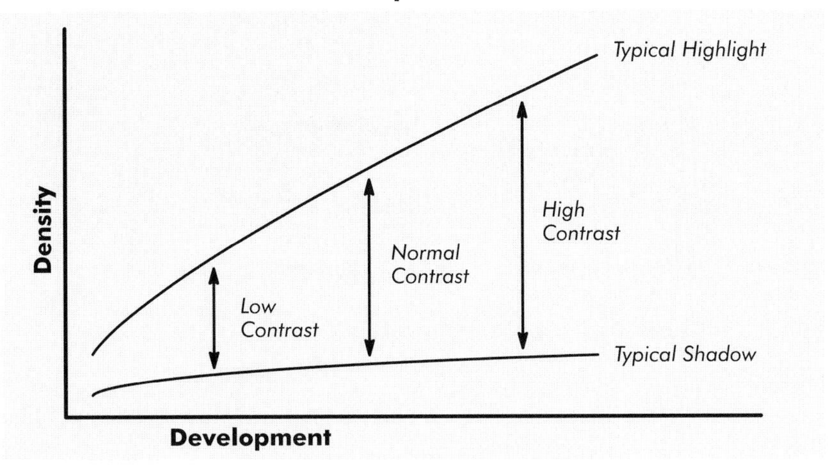

This graph illustrates the effect of increasing development on contrast. The curves representing a typical highlight density and a typical shadow density show that as development increases, the separation between the highlight and shadow (contrast) increases.

How Previsualization Affects Contrast

The previous graph also illustrates how you use the zone system. First, you use placement to "expose for the shadows," and then you "develop for the highlights" to achieve the amount of contrast you want in the negative.

At one point during development, both the highlight and shadow densities will produce the tones on the print that you previsualized. Before that, the highlight and shadow densities are too close together for your previsualization, and the image is low in contrast, or flat. With prolonged development, there is too much separation between highlights and shadows. The result is higher contrast than your previsualization.

Print from Underdeveloped Negative

Print from Correctly Developed Negative

Print from Overdeveloped Negative

The three prints above illustrate how contrast increases as you increase film development. There is only one point during development where the negative densities will produce the previsualized tones. When the gray card is printed to match a zone V tone, the underdeveloped negative makes the white pants (zone VII) appear too dark and the black drape (zone III) appear too light. Using the correctly developed negative, the white pants and the black drape fit the previsualization of the image. Using the overdeveloped negative, the white pants appear too light and the drape appears too dark.

Because there are many possible ways to previsualize a scene, there can be no single development time for film. Each time you previsualize a scene, whether the contrast is N, N+1, or N−2, you are specifying a different development time for the film, one that will develop the highlights to the density that fits your needs for the image. In black-and-white photography, each film and developer combination is different, so that the only way to discover the actual time for a specific contrast is through testing. This is the subject of Chapter 9, *Finding Your Development Times*.

MECHANICAL CONTROL OR CREATIVITY?

In some ways, nineteenth-century photographers were correct in thinking that the final step in photographic creativity is in the development process. But they were wrong in assuming that changing the image with development was subject to the whims of the photographer.

Hurter and Driffield proved that density changes and contrast were a function of exposure and development in a completely predictable process. This came as a great blow to photographic artists. It reduced to simple cause and effect what to them had been artistic prerogatives. The photograph now seemed to rely too much on mechanical devices and chemical baths rather than on the kind of pure and direct communication that a painter has with the canvas.

A champion of photography in the nineteenth century, a doctor named Peter Henry Emerson, even went so far as to publicly denounce the idea that photography could any longer be considered an art form when he heard about Hurter and Driffield's experiments. Interestingly, however, photography did not change as a result of Emerson's dissatisfaction. In fact, even while Emerson was declaring that "the medium [of photography] must always rank the lowest of all the arts,"* he continued to make the same sensitive images of rural England that he had made before.

There is an important lesson in this. Before Hurter and Driffield's discovery, photography was a creative medium only because photographers thought they could randomly apply technique to produce a desired result. Since that time, photographers have come to the realization that only through the understanding and application of careful and predictable craft can you be free to visually express feelings in harmony with the materials. As with any expressive form, photographic materials require a discipline that a photographer must understand before using them effectively.

* Newhall, *op. cit.*, p. 145.

SUMMARY

The contrast of a scene is a combination of previsualization and measuring light reflectance. The contrast of a negative is determined by how you develop the film.

- The amount of exposure determines the shadow densities in a negative. No amount of additional development can add significant density to a shadow tone.
- A highlight tone builds up density in a negative as development increases.
- The more you develop film, the greater the difference in density between highlights and shadows. This separation of tones is the definition of contrast in a negative.
- The zone system adage, "expose for the shadows and develop for the highlights," refers to placing shadows and developing for the contrast that you previsualize.
- There is only one development time that will produce the print tones that you previsualize. Every previsualization has a potentially different development time.
- The fact that photography is a product of specific physical and chemical processes does not mean that it cannot be a creative medium. Being a creative photographer is a matter of understanding the nature of the materials and working in harmony with them.

Chapter 7

Making the Zone System Work

Directional sunlight and deep shadows are often difficult tones to control. Careful previsualization, exposure placement, and contrast determination are the tools you need to bring these picture elements under control.

Understanding the theory of the zone system is not the same as mastering its application. This chapter takes you through the process of actually using the zone system to make photographs, first in a general way, and then using two specific examples. As the chapter progresses, you will learn some techniques for using the zone system in unusual or difficult lighting situations.

PUTTING THE PUZZLE TOGETHER

Knowing the zones and understanding how exposure and development affect film density are the pieces of the zone system puzzle. Making these pieces fit together with your previsualization is the next step. As you have read in previous chapters, the steps are as follows:

- First, decide how you want the final print of a scene to look through previsualization.

- Next, take indicated meter readings of the light reflected from the areas you previsualize.
- Finally, use your indicated meter readings to determine the exposure placement and contrast of the scene.

The following sections discuss each of these steps in greater detail.

Previsualizing a Scene

In theory, previsualization means establishing a mental image of the final print before you expose the film. It is a way to see the scene in the same shades of black, white, and gray that the film and printing paper see. But, given all the available choices for previsualizing most scenes, it can be confusing to decide which shadow zone to place and which highlight zone to compare it to when you are determining contrast.

In general, the farther apart the zones you choose for the most important shadow and most important highlight, the more accurate your previsualization will be. Zones on the extreme ends of the zone system scale (zones I and IX), however, are usually difficult to accurately read with a light meter and relate to the zones in the middle of the scale. It is the zones with detail that are easiest to previsualize in most scenes and the ones you should concentrate on.

Zones With Detail

Zones with detail and texture are easier to previsualize than zones without. Zones III and IV, for example, are usually the easiest shadow zones to previsualize. In practically every scene there should be an area that you want to appear as dark as possible in the print and still show detail (zone III), or an area that you want to appear as a dark but open and fully detailed shadow (zone IV).

Two highlight zones that are easier to previsualize than other highlights are zone VII and zone VI. Zone VII is the area that you want to appear as light as possible on the print and still show texture. You can compare zone VI to the clear north sky or Caucasian skin. Most scenes contain highlights that you can identify as either zone VII or zone VI.

In all cases, you should avoid previsualizing zones I and IX as the most important shadow and highlight in a scene. These zones are so extreme and are usually such small areas of an image that they can make accurate exposure and contrast determination difficult. This is not to say that these zones cannot be important to the image, merely that they are not generally helpful in your zone system calculations.

Difficult Scenes

Occasionally, even experienced zone system photographers have difficulty deciding which areas in a scene are the most important shadows and highlights. This can especially be a problem when the contrast in a scene is low. Frequently, in such scenes there are no distinct areas that stand out as obvious candidates for shadows or highlights with detail and texture.

When you encounter a hard to previsualize scene, try squinting your eyes so that the image before you appears out of focus. When you are not distracted by details, broad areas of tone, including the highlights and shadows, become more noticeable.

These two images illustrate how throwing your eyes out of focus can cause the most important highlights and shadows in a hard-to-previsualize scene to become more obvious. Without the distraction of detail, you can concentrate on the tones.

Another technique that you can use in difficult situations is to compare indicated meter readings of all the areas that you suspect might be shadows or highlights. This allows you to mentally juggle the tones in the image before deciding which ones are the most important for exposure and contrast. See the next section, *Taking Indicated Meter Readings*, for examples of how to use this technique.

Taking Indicated Meter Readings

Once you previsualize a scene, measure the important highlights and shadows with your light meter. Make a note of your indicated meter readings. It helps to carry a notebook with you when you are starting out so that you can actually write the readings down.

If you feel confident about your previsualization, you only need to take meter readings of one shadow and one highlight area. If you are not completely sure, then you should measure as many possible shadows and highlights as you need to support your previsualization. Remember that an indicated meter reading is the exposure that will render the particular tone you are measuring as a medium gray (zone V) density on the film.

Changing Your Previsualization

When you compare indicated meter readings of certain scenes, you may want to change your previsualization. Knowing the different possibilities for how shadows and highlights can appear in the final print allows you to play "what if" games with your previsualization and to accurately predict the result of any possible scenario.

For example, a scene in which three potential zone III shadows all have different indicated meter readings might cause you to change how you place the exposure. If you place the shadow reflecting the least light in zone III, the other two will appear lighter in the print. If one of the remaining shadows reflects a full stop more light than the one you place, it will appear a full zone lighter, in zone IV. On the other hand, placing the shadow reflecting the most light in zone III will make the other two shadows appear darker, losing detail as they fall lower than zone III.

A similar situation can occur when you compare indicated meter readings of highlights. A scene in which three potential zone VII highlights all have different indicated meter readings might have different contrasts depending on which highlight you choose as the zone VII. If one of the potential zone VII highlights reflects a full stop more light than the others, choosing it over one of the less reflective highlights will change the contrast of the scene. What might be a normal contrast scene (N) if you chose the less reflective highlight becomes a high contrast scene (N−1) when you choose the most reflective highlight as the zone VII.

Exposure Placement

Your exposure for a scene depends on where you choose to place the most important shadow. You can have a shadow appear in the print as a dark detailed shadow (zone III), an open shadow (zone IV), or a very dark shadow with little or no detail (zone II). Nothing you do later in developing the film or making the print can undo the shadow placement.

Place the shadow by taking its indicated meter reading, and stop down one f-stop (or shutter speed) for each zone your previsualization is below zone V. For example, if your previsualization is zone IV and the indicated meter reading is f/8 at $^1/_{60}$, then you place the shadow in zone IV by stopping down to f/11 at $^1/_{60}$ (or f/8 at $^1/_{125}$). To place the shadow in zone III, you stop down two stops, to f/8 at $^1/_{250}$ (or its equivalent). To place the shadow in zone II requires an exposure of f/16 at $^1/_{250}$ (or its equivalent).

Alternative Placement versus "Bracketing"

Once you are comfortable with how placement affects the shadows, you can use alternative previsualization and placement as a way to experiment with exposures. In the past, where you might have been tempted to "bracket" your exposure (one stop less and one stop more than your average meter reading) in hopes that you would get at least one printable negative, you can now practice alternative placement to explore the effect of different exposures on the shadow tones.

Although alternative placement and bracketing can sometimes yield the same results, bracketing is a virtually random practice, while choosing different exposures based on alternative shadow placements is a way to explore what happens when you consciously previsualize the important shadow in different zones. An alternative placement, for example, lets you see the differences between a zone II shadow (no real detail), a zone III shadow (some detail), or a zone IV shadow (great detail) in your image.

Another benefit of using alternative placement is that they can confirm your equipment calibration for the correct exposure (discussed in the next chapter). The different exposure placements let you see if you are getting the correct shadow density on the film. If your placed zone III does not show any detail in the negative, it indicates that the film is underexposed. If a placed zone II has detail, then that indicates that the film is overexposed.

Contrast Determination

Determining the most important highlight in a scene affects the contrast of that scene. Once you choose zones for the most important highlight and shadow, compare their indicated meter readings. Calculate the difference in stops between the two readings to the difference in zones of your previsualization.

For example, consider a scene in which you previsualize the most important shadow in zone III and the most important highlight in zone VII. Your indicated meter reading for the shadow is f/8 at $^1/_{60}$, and your indicated meter reading for the highlight is f/22 at $^1/_{125}$. The difference in stops between the two indicated meter readings is four. The difference in zones between zone III and zone VII is also four. The result is a normal contrast scene (N).

Changing your previsualization, however, changes the contrast of a scene, even if the indicated meter readings stay the same. If you previsualize the highlight in the first example as a zone VI instead of a zone VII, the difference in zones between the shadow and highlight is now three. Since the difference between indicated meter readings is still four stops, that means you now have a high contrast scene (N−1). The difference in the two examples is one of previsualization, not light reflection.

MAKING THE ZONE SYSTEM WORK — TWO EXAMPLES

The following examples describe how to apply the zone system in two typical situations, a landscape and a portrait. Each example discusses previsualization, metering, exposure placement, and contrast determination.

Example One: A Landscape

The landscape in the first example is typical of an image with tones you can previsualize in a number of different ways and still produce a print that most viewers will consider to be an accurate representation of the scene. The following sections illustrate two ways of previsualizing and printing this scene.

As you approach the scene you want to photograph, decide how you want the major tonal areas in the scene to appear in the final print. Assign zones to these areas.

Previsualization

You decide that the ferns in the foreground are middle-gray tones and the area behind them is the darkest part of the image, though it appears to be open shade. You previsualize the ferns as a zone V and the darkest part of the background as zone IV. The most important highlights are on the tree trunk. You decide that the bark must have texture and detail, so you previsualize it as a zone VII. The previsualization of the most brightly lit parts of the trunk does not matter as long as it is very light. It can be either a zone VIII or a zone IX.

If you are using an in-camera light meter, walk up close to each previsualized area to make your indicated meter readings.

Indicated Meter Readings

Next, take meter reading from the ferns, the shaded background, and the tree trunk. Although an indicated meter reading from the ferns is not strictly necessary since you previsualized them as a midtone (zone V), knowing how much light they reflect compared to the more important highlights and shadows lets you to visualize exactly how they will appear in the final print.

Object	Previsualized Zone	Indicated Meter Reading
Background	Zone IV (open shade)	f/5.6 at 1/60
Ferns	Zone V (midtone)	f/5.6 at 1/125
Trunk	Zone VII (detailed highlight)	f/8 at 1/250
Sunlit trunk	Zone VIII or IX (bright highlight)	f/11 at 1/500

A sketch of the major areas of the scene, with notes indicating the previsualization and the indicated meter readings, can help you decide how to place the exposure and to determine the contrast.

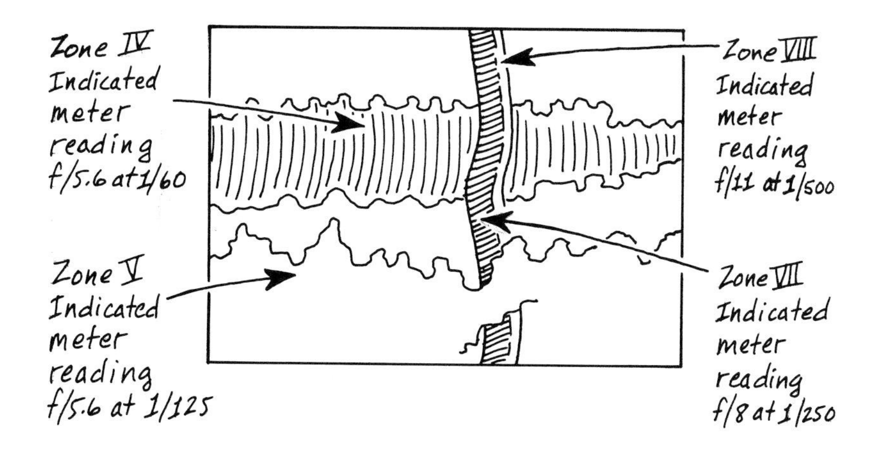

Carrying a notebook lets you sketch the important areas of the scene and add your previsualization and the indicated meter readings. This is a useful tool, especially as you are learning the zone system, or when you encounter particularly difficult scenes.

Exposure Placement

Placing the shaded background in zone IV means taking the indicated meter reading of f/5.6 at $^1/_{60}$ and stopping down one stop. For this scene, use the exposure f/8 at $^1/_{60}$ to enhance the depth of field instead of f/5.6 at $^1/_{125}$. The landscape benefits from a greater overall appearance of sharpness and does not have any moving objects that might require the faster shutter speed. Use a tripod if you are not sure that you can hold the camera steady at $^1/_{60}$ of a second.

Contrast Determination

Compare the indicated meter readings for the shaded background (zone IV) and the tree trunk (zone VII). They are f/5.6 at $^1/_{60}$ and f/8 at $^1/_{250}$ (three stops apart). As this is one stop of exposure difference for each previsualized zone, the contrast is normal (N).

The final print shows the tones in the scene as you previsualized them. Notice how the ferns appear in zone V and the tree trunk in zone VII once the exposure placed the background into zone IV.

The Final Image

The print (on normal-contrast printing paper) has tones that appear the way you previsualized them. The shadow in the background is highly detailed, appearing where it was placed, in zone IV. The ferns in the foreground are medium gray (zone V), appearing one zone lighter than the background. The majority of the tree trunk appears three zones lighter (zone VII) than the placed shadow, with the exception of the most brightly lit portion, which appears in zone IX, five zones lighter.

Alternate Previsualization

The first example illustrates one way to previsualize this scene. Another method would be to previsualize the background as zone III and keep the tree trunk as zone VII. Assuming that the indicated meter readings remain the same, the exposure placement would now be f/11 at $^1/_{60}$ (two stops less than the indicated meter reading of the shadow). Because the difference in zones is four, the contrast is now N−1. This makes the background and the ferns darker than they appeared in the first example.

This alternative previsualization of the landscape example shows how a one zone darker previsualization of the background makes the highlights in the tree trunk appear brighter in the image. To emphasize the highlights, the image is composed as a vertical. This shows how previsualization and composition can mutually support each other.

Example Two: A Portrait

The second example is a portrait. Compared with the landscape, the options for previsualizing skin tone are more limited if you want to make an image that is about the people in the photograph and not the tones.

Previsualization

You decide to position the subjects against partially shaded foliage for a background. The darkest foliage just behind the subjects becomes the most important shadow area. Because it is open shade, you previsualize it as zone IV. The faces are zone VI, the representational previsualization of Caucasian skin. The white shirt on the child is an important highlight, but whether or not it appears in the print as a zone VII or a zone VIII is not as critical as having the skin appear in zone VI.

For this image, the shaded foliage is the most important shadow and the skin tone is the most important highlight. The white shirt is a highlight that can appear either in zone VII or zone VIII.

Indicated Meter Readings

Take meter readings from the shaded foliage, the faces of the subjects, and the white shirt. You can also take a meter reading from the grass in the foreground. Although an indicated meter reading from the grass is not necessary, knowing how much light it reflects gives you a reading to compare with the previsualized zone IV foliage and the zone VI skin.

With willing subjects, you can walk up close to the previsualized tones to take meter readings.

The following table shows the indicated meter readings from the previsualized areas:

Object	Previsualized Zone	Indicated Meter Reading
Background	Zone IV (open shade)	f/5.6 at $^{1}/_{60}$
Face	Zone VI (highlight)	f/8 at $^{1}/_{250}$
White Shirt	Zone VII or VIII (bright highlight)	f/11 at $^{1}/_{500}$

The following sketch indicates the major areas of the scene with notes indicating the previsualization and the indicated meter readings of these areas.

The sketch gives you a record of your intentions for the image with which you can compare the final result. It is better not to rely on memory when you are shooting several images at a time or when it might be days before you are able to print the negative.

Exposure Placement

Place the shaded foliage in zone IV by stopping the indicated meter reading of f/5.6 at $^{1}/_{60}$ down one stop. For this scene, use the exposure f/5.6 at $^{1}/_{125}$. The higher shutter speed will prevent any unanticipated movement by the young child from causing a blur.

Contrast Determination

Compare the indicated meter readings for the shaded foliage (the previsualized zone IV) and skin tone (the previsualized zone VI). They are three stops apart, while their previsualization is two zones apart. As this is one stop greater than the difference in previsualized zones, the contrast is high (N−1).

The final image (on normal contrast printing paper) shows the tones as you previsualized them. The background foliage is in zone IV (open shadow), the skin, in zone VI, and the white shirt ended up in zone VIII.

Alternate Previsualizations

When photographing people, it is difficult to choose a previsualization that alters the skin tone. Since the most important highlight in this scene is the zone VI skin tone, there are no alternative previsualizations that would make a convincing portrait. You could choose to alter the rendering of these tones, but the result would be less of a portrait and more of an abstraction.

SUMMARY

Exposure and development both have a specific effect on film density. Combining their effects with your vision is the secret to using the zone system.

- Follow these steps when using the zone system: Previsualize the scene, measure the light reflected from the important shadows and highlights, then use the indicated meter readings to determine exposure placement and contrast.
- When deciding what to previsualize in a scene, try to choose highlights and shadows that are as far apart on the zone system scale as possible but not at the extremes where accurate metering is difficult.
- The darkest and lightest zones with detail and texture, zone III and zone VII, are among the easiest to previsualize accurately. Other shadows and highlights with detail, such as zone IV and zone VI, are also relatively easy to previsualize.
- In previsualizing a difficult scene, it can help to throw your eyes out of focus so that you see the broad areas of tone in the scene without being distracted by detail.
- Use indicated meter readings from different shadows and highlights to mentally juggle your previsualization of a scene and still predict what the final print will look like.
- Trying alternative previsualizations of the shadows in a scene is a way to use different exposures to experiment with different shadow renderings. It is also a way to confirm the accuracy of your exposures.
- Changing your previsualization often changes the contrast of a scene by changing the relationship between the previsualized zones and the indicated meter readings of those zones.

Chapter **8**

Finding Your Exposure Index

Using the correct exposure index means that the zones you place will appear on the print as you previsualize them. If, for example, the black stripes on this lighthouse had been a zone lighter or darker than the zone III in which they were placed, the sense of darkness and detail would have been lost.

To successfully use the zone system, you must be sure that your equipment and working procedures can produce the tones you previsualize. In previous chapters, you learned that correct exposure and film development are the most important variables that affect film density. Since a correct exposure is necessary before you can find your correct development times, the first calibration test is to establish an exposure index.

WHAT AN EXPOSURE INDEX IS

An exposure index is the calibration you use to compensate for possible errors in your camera's shutter or light meter, and to take into account the effect of a specific film and developer. Accurate. exposures using the zone system depend on your ability to place a previsualized shadow zone. But correct exposure placement is impossible if your camera and light meter do not produce the correct film density. Such a failure is almost always caused by a problem with the light meter, the camera shutter, or frequently, with both.

Equipment error is the result of a number of factors. Cameras and light meters rarely leave the factory perfectly calibrated. Even in cases where a manufacturer's tolerance for assembly error is very small, subsequent shipping and handling can knock the calibration off. The solution for these and other problems is to first establish that any error is a consistent one, and then to modify the way you expose film to compensate for the error.

Light Meter Errors

The standard light meter calibration is for a gray tone that reflects 18% of the light that falls on it. In the zone system, this is zone V. It is not uncommon, however, for light meters to actually produce readings as much as one stop away from the 18% reflectance standard. If the error is on the plus side, a one stop error would make a zone III placement a zone IV (too much density and extra grain in the image). Even worse, if the error is on the minus side, a zone III placement would actually be a zone II that has no detail.

The 18% reflectance value of the Kodak Neutral Test Card (gray card) is the standard to which all light meters are calibrated. Many meters vary from this standard, however, either because of wear or manufacturing error.

After manufacturing and shipping, two common causes of light meter problems are the battery and careless use. If your meter is a battery resistance type, then the age and voltage of the battery can cause incorrect readings. The cure is prevention. Check the battery regularly, and replace it at least once a year. If you are unsure about the condition of your battery now, replace it before you begin the exposure index test.

Another problem is mishandling. Pointing a camera with a built-in meter at the sun while the meter is turned on, is a common way to get underexposures. Many older light meter cells have a memory and are "blinded" by the light of the sun for up to 15 minutes. Memory also affects readings when you store a camera in the dark for long periods of time with the meter turned on. For several minutes after you bring it into the light, the readings will tend toward overexposure. Most light meters currently manufactured are designed to minimize these problems, but in extreme situations they can still be a problem.

Shutter Errors

Camera shutters also have calibration problems. Coming off the assembly line, no two mechanically driven shutter mechanisms are exactly alike and, as they are used, the gears gradually wear, springs lose their tension, and lubricant dries out and gets redistributed. Even the newer electronic shutters, while generally more accurate than the purely mechanical ones, can give exposure times as much as 50% different from their set times. When a shutter error is compounded by a light meter error, your exposures can be more than one stop off.

One way to slow down wear on a mechanical shutter is to store your camera with the shutter uncocked. This relaxes the tension on the springs that drive the mechanism. Long periods of tension cause the springs to weaken and the timing to become erratic. Even if you leave the shutter uncocked, it is a good idea to "exercise" your camera if you have had it in storage for a while. Before you load any film, rapidly cock and release the shutter five or six times at each shutter speed. Electronic shutters are less prone to problems than mechanical shutters but require fresh batteries in order to maintain their accuracy.

COMPENSATING FOR ERRORS

No two cameras are exactly the same. Every camera has a different combination of errors. Some cameras may even produce good exposures without any compensation. This might occur for many reasons, including errors that cancel each other out; for example, a fast shutter may cancel the error of a light meter that reads too low. Even if your light meter and shutter are in adjustment, that does not necessarily take into account your film and developer combination or your working methods.

One fortunate thing about the inevitability of errors is their reliability. Unless the cause is serious, most exposure errors are consistent. Light meters generally read consistently above or below the 18% reflectance standard, and most out-of-calibration shutters are consistently slow or fast.

When an exposure error is consistent, you can compensate for that error by changing your light meter setting. Regardless of their source, all exposure errors affect the final negative density, and your light meter is the easiest component in the chain to recalibrate. You recalibrate a light meter by simply changing the setting on the film speed dial.

Film speeds are rated by ISO numbers, though on older cameras and light meters, this dial may have ASA or even DIN numbers.* ISO numbers double or halve as a film's sensitivity to light increases or decreases by a stop.

* The International Standards Organization (ISO) is the group that determines many standards, including those for film speed. ISO film speed ratings are equivalent to the older ASA (American Standards Association) ratings. The European standard, DIN, uses a different numbering system.

Exposure Index Is Different from Film Speed

Normally, the ISO dial on a light meter refers only to the light sensitivity of the film in the camera. When you use this dial to compensate for your equipment or procedures, however, you call the setting an *exposure index*, or *E.I.* So, while Kodak's 35mm Tri-X film has an ISO rating of 400, you may have to set your camera's light meter dial to a higher or lower number in order to get accurate exposures with it. Calling this number an exposure index instead of ISO is not just a semantic difference. The term *exposure index* makes you aware that you are taking into account the interdependence of the entire exposing system — light meter and shutter — as well as the film's sensitivity to light.

THE EXPOSURE INDEX TEST

The test to find your exposure index in this chapter is an empirical one. Although you could use various devices that measure film density to find the same information, this test duplicates as much as possible how you use your camera to take pictures. This is what is meant by empirical; the test relies on practical experience rather than laboratory experiments. You do not need any special equipment, only your camera and a standard black-and-white darkroom.

If you have more than one camera body, you must test each one separately because the shutter mechanisms in most 35mm cameras usually vary from body to body. On the other hand, you should not have to test each of your camera's lenses. Most aperture settings are consistent from lens to lens, especially when they are the same brand.

Materials

You need the following materials for the exposure index test:

- a 36-exposure roll of the film that you plan to test
- an 18% gray card (Kodak or other brand)
- an object with zone III detail, such as a dark blue or black knitted sweater
- an object with zone VII texture, such as a white sweater or a white terry cloth towel
- a willing and patient subject

A tripod is useful, but optional.

Choosing Your Film and Developer

You are calibrating the materials that you use as much as your camera shutter and light meter when you perform the zone system tests in this book. Changing film or developer requires an entirely new set of calibration tests. Even the type of enlarging paper that you use makes a difference. You should carefully select the film, film developer, and enlarging paper for your tests, and then use them consistently.

In selecting black-and-white films, the basic choices are in emulsion speed. The important factors to consider are the grain size, which is finer in the slower speed films, and the latitude for controlling contrast, which is greater in higher speed films. Brand names are unimportant, except for their availability in your area.

Developers, too, have certain basic characteristics. Standard developers, such as Kodak's D-76 and Ilford's ID-11, offer fine grain and fairly flexible contrast control. Ultra-fine grain developers, such as Kodak's Microdol-X and Edwal's FG-7 with added sodium sulfite, offer the appearance of smoother grain at the sacrifice of image sharpness (the technical term is *acutance*). High acutance developers offer the sharpest possible image, but also make grain appear more clearly defined. Examples of such developers are Beseler's Ultrafin and Agfa's Rodinal (when used at high dilutions).

The combination of film and developer is important. Each combination produces an image with a different character. It takes time and consistent use of a film and developer to learn these qualities. Switching back and forth between films and developers can slow down the learning process.

Enlarging paper also has certain characteristics that affect your prints. Normal contrast grades of different brands of paper can produce slightly different renderings of tones. If you do your tests on one brand of paper, expect some changes if you switch to another type of paper. The differences may be subtle, but, as you will notice, one of the benefits of performing the tests in this book is to increase your perception of these differences.

Preparation

1. Take your subject outside into normal contrast light, such as the north side of a house on a sunny day. The location can be anywhere that your light meter indicates a four-stop difference between the zone III and zone VII objects that you selected for this test. Even though you have not yet calibrated the light meter, the comparison between indicated meter readings should be accurate.
2. Stand your subject against a neutral background. Have the subject hold the gray card while wearing (or while draped with) the zone III and zone VII objects.
3. Compose a "head and shoulder" portrait in your viewfinder, filling the frame with as much of your subject's face as possible and still showing the zone III object, the zone VII object, and the zone V gray card in the frame. Use this composition for the whole test. You may want to use a tripod, but a tripod is not necessary unless your exposure times are longer than 1/60 of a second.
4. Set the dial on your light meter to the film's ISO rating. Because all the exposures you make on this roll are tests to find the correct setting for your exposure index, the procedures will now refer to the film speed numbers on the dial as the *E.I.*
5. Meter the zone III object. If your camera has a built-in light meter, be sure to follow the procedure and cautions described in the section *Using a Camera's Built-in Light Meter* in Chapter 4.
6. Place the indicated meter reading in zone III by stopping down two stops.

Keep careful notes of all your indicated meter readings and exposure placements throughout the test. Use the form on page 67 for your data. Should something go wrong and show up later in the test, your notes will prove invaluable in locating the problem. Never trust your memory. An added advantage of recording this information as you shoot is that it forces you to work more slowly and minimizes careless errors.

For the exposure index test, compose a "head-and-shoulders" portrait that includes the previsualized highlight, shadow, gray card, and your subject's face.

Procedure

1. FRAME 1: Expose at the same E.I. as the film's ISO rating.
2. FRAME 2: Cover the lens and shoot a blank frame. Although it is not absolutely necessary, the blank frame will help to identify each test frame after you develop the film.
3. FRAME 3: Use an E.I. that is four times the film's ISO rating. Re-meter the previsualized zone III and place it exactly as you did for the first frame (stop down two stops). Make a note of your indicated meter reading and exposure placement. If the light has not changed from the time you metered the first frame, then the exposure placement should be two stops less. If it is not, that indicates either a procedural error or an inconsistency in the meter that needs repairing. First, check your procedures.
4. FRAME 4: Blank frame.
5. FRAME 5: Use an E.I. that is two times the film's ISO rating. Re-meter the zone III and place it as in frame 3 (stop down two stops).
6. FRAME 6: Blank frame.
7. FRAME 7: Use an E.I. that is the same as the film's ISO rating. The exposure should be the same as frame 1. This is another opportunity to check your procedures and the consistency of your light meter.
8. FRAME 8: Blank frame.
9. FRAME 9: Use an E.I. that is $1/2$ the film's ISO rating.
10. FRAME 10: Blank frame.
11. FRAME 11: Use an E.I. that is $1/4$ the film's ISO rating.

12 FRAME 12: Blank frame.
13 FRAME 13: Use an E.I. that is 1/8 the film's ISO rating.
14 FRAME 14: Blank frame.
15 FRAME 15: Use an E.I. that is 1/16 the film's ISO rating
16 FRAME 16: Blank frame.

Checking Your Notes

At frame 16, stop and study your notes. If you are testing a fast black-and-white film (ISO 400), then your test E.I.s should be 400 (frames 1 and 7), 1600 (frame 3), 800 (frame 5), 200 (frame 9), 100 (frame 11), 50 (frame 13), and 25 (frame 15). Films with other ISO ratings should have a similar series of test exposure index settings based on their emulsion speeds.

Your notes should also indicate a relationship between each exposure of one stop, except between frames 1 and 3, which should be two stops apart. Only changing light, such as that found on a partly cloudy day, is an acceptable reason for a departure from this relationship. Otherwise, use the remainder of the roll to shoot the test over.

Final Exposures

If the information in your notes looks correct, reset your camera's meter to an exposure index that is the same as the film's ISO speed rating and shoot the rest of the roll in the following manner:

- First, find a scene that you want to photograph and previsualize a significant shadow in it. Place this shadow in the previsualized zone (zone II, III, or IV) and expose the frame.
- Next, expose the same scene one stop more and one stop less than the placed exposure. This is the equivalent of testing three different exposure indexes: the same as the film's ISO, half the film's ISO, and twice the film's ISO.
- Finish the rest of the roll in this manner, choosing scenes that interest you. Expose the first frame for a placed shadow zone. Then expose two more frames, one a stop more and one a stop less than the initial exposure placement.

The contrast does not matter in these final scenes, but this is an opportunity to practice metering both highlights and shadows, and to make notes about the contrast as well as the exposure placement.

Film Processing

After you expose the test roll of film, process it as you normally would using the developer of your choice. Use the film manufacturer's recommendations for time and temperature if you do not have any other information.

Be sure that you agitate the film at a consistent rate during development. Consistency is more important than the agitation technique you use. Normally, you should agitate constantly for the first minute and then for 10 seconds every minute after the first.

Pages 66 and 67 illustrate a form you can use to complete the exposure index test. The form contains an outline of the procedures as well as a place to record the information you need to evaluate your negatives and to organize your results clearly. Page 66 contains the data recorded from the exposure index test used as the illustration for this chapter. Page 67 is left blank so you can photocopy it for your personal use.

EXPOSURE RECORD E.I Test

DATE 10-3-96 FILM USED Kodak Tri-X (ISO 400)

CAMERA USED Nikon - Ser. # 7253119

COMMENTS Previsualized hair as Zone III, sweater as Zone IV, skin tone (face) as Zone VI, blouse as Zone VII, indicated meter readings (w/meter set at film's ISO) were f/4 at 1/125 sec. for hair and f/16 at 1/125 sec. for blouse, -normal contrast scene-

Frame	Procedures	Indicated meter reading off zone III shadow	Calculated exposure (placed in zone III)
1	Expose at same E.I. as the film's ISO	f/4 at 1/125 sec.	f/8 at 1/125 sec.
2	Cover the lens and shoot a blank		
3	Set the E.I. to 4 times the film's ISO	f/8 at 1/125 sec.	f/16 at 1/125 sec.
4	Shoot a blank		
5	E.I. at 2 times the film's ISO	f/5.6 at 1/125 sec.	f/11 at 1/125 sec.
6	Shoot a blank		
7	E.I. at the same rating as the film's ISO	f/4 at 1/125 sec.	f/8 at 1/125 sec.
8	Shoot a blank		
9	E.I. at 1/2 the film's ISO	f/2.8 at 1/125 sec.	f/5.6 at 1/125 sec.
10	Shoot a blank		
11	E.I. at 1/4 the film's ISO	f/2.8 at 1/60 sec.	f/5.6 at 1/60 sec.
12	Shoot a blank		
13	E.I. at 1/8 the film's ISO	f/2.8 at 1/30 sec.	f/5.6 at 1/30 sec.*
14	Shoot a blank		
15	E.I. at 1/16 the film's ISO	f/2.8 at 1/15 sec.	f/5.6 at 1/15 sec.*
16	Shoot a blank		
	Review your notes. Re-shoot any exposures that look incorrect. It is easier to do it now than to start all over again.		
17	Re-shoot E.I. ________		
18	Shoot a blank	Exposures look OK.	
19	Re-shoot E.I. ________		
20	Shoot a blank		
19	Re-shoot E.I. ________		
20	Shoot a blank	* Note: Used tripod for these frames.	

EXPOSURE RECORD ____

DATE ____________________ FILM USED ____________________

CAMERA USED ____

COMMENTS ____

____

Frame	Procedures	Indicated meter reading off zone III shadow	Calculated exposure (placed in zone III)
1	Expose at same E.I. as the film's ISO		
2	Cover the lens and shoot a blank		
3	Set the E.I. to 4 times the film's ISO		
4	Shoot a blank		
5	E.I. at 2 times the film's ISO		
6	Shoot a blank		
7	E.I. at the same rating as the film's ISO		
8	Shoot a blank		
9	E.I. at $^1/_2$ the film's ISO		
10	Shoot a blank		
11	E.I. at $^1/_4$ the film's ISO		
12	Shoot a blank		
13	E.I. at $^1/_8$ the film's ISO		
14	Shoot a blank		
15	E.I. at $^1/_{16}$ the film's ISO		
16	Shoot a blank		
	Review your notes. Re-shoot any exposures that look incorrect. It is easier to do it now than to start all over again.		
17	Re-shoot E.I. ________		
18	Shoot a blank		
19	Re-shoot E.I. ________		
20	Shoot a blank		
19	Re-shoot E.I. ________		
20	Shoot a blank		

Constant temperature is also important. Be sure that you have an accurate thermometer. Dial-type and inexpensive digital thermometers are rarely accurate. If you use one, check it periodically against a glass mercury thermometer, such as those sold for color processing, and compensate for any inaccuracy.

Once you have developed the film from the exposure index test, examine it carefully and identify the index you used for each frame. What you see in the negatives at this point will be the first indication of which negative is correctly exposed.

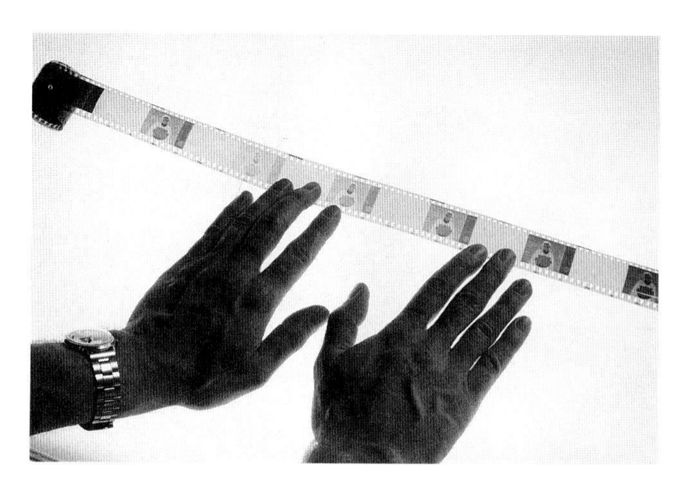

Inspecting the Negatives

After the processed film is dry, carefully examine the negatives using the following steps as a guide. A magnifier and an even source of light will help. Page 69 shows the negatives from the exposure index test used as an illustration in this chapter.

1 Arrange the negatives in order from the least exposure (highest E.I.) to the greatest exposure (lowest E.I.) Remember that frame 1 and frame 7 are duplicate exposures.
2 Look at the zone III shadow in each negative. Compare frames 1 and 3. Frame 3, which you exposed two stops less than frame 1, should lack detail in the previsualized zone III area.
3 Continue looking at the other frames in sequence until you find the first frame that has detail in the zone III area. The first frame to show a proper zone III density is usually the frame exposed at the correct E.I. This is just a preliminary indication, however. A negative can have faintly visible detail that will not show up in the print.
4 If no zone III detail appears in frame 1, then do not expect detail to show up until at least frame 9. This is an indication that your E.I. is less than the film's ISO rating.
5 If detail appears in frame 3, it indicates that your E.I. is much higher than the film's ISO rating. If this is the case, you must shoot a new test at E.I.s that are 8 times and 16 times the film's rated ISO to be sure that you have an accurate test.

Frame 3: E.I. 1600

Frame 5: E.I. 800

Frame 7: E.I. 400

Frame 9: E.I. 200

Frame 11: E.I. 100

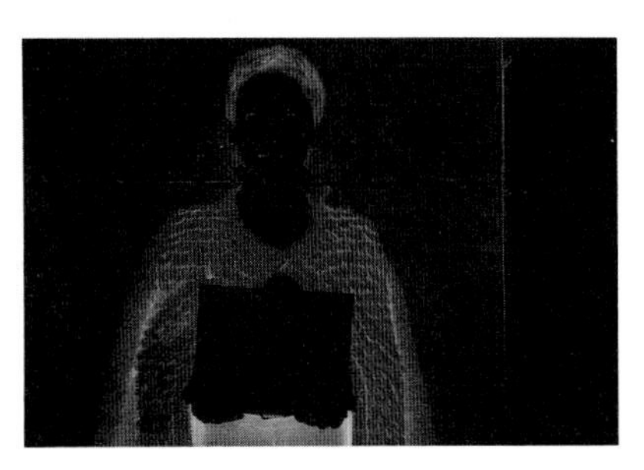

Frame 13: E.I. 50

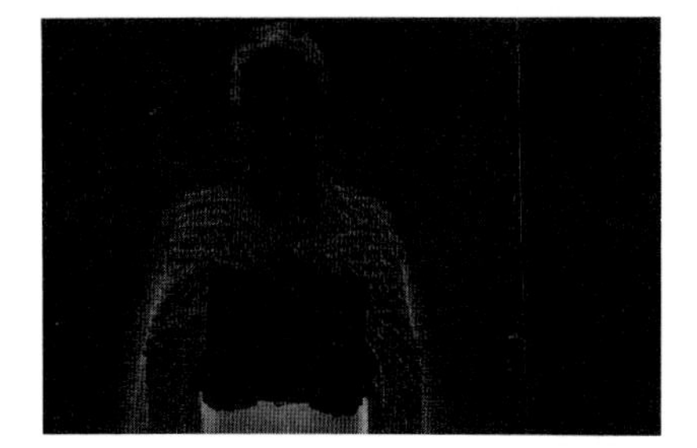

Frame 15: E.I. 25

As you examine the processed film from the exposure index test, identify the index of each test frame. What you see in the negatives at this point is the first indication of the correct exposure index. In this series of negatives, frames 3 (E.I. 1600) and 5 (E.I. 800) do not show any detail in the hair (the previsualized and placed zone III value). Frames 7 (E.I. 400) and 9 (E.I. 200) show adequate detail and indicate that the correct index is probably in the range of 400 to 200. Frames 11 (E.I. 100), 13 (E.I. 50), and 15 (E.I. 25) show detail in the zone III area, but also have extra density, which will show up as increased grain on the print.

The Contact Sheet (Proper Proof)

Once you examine the negatives, make a contact sheet of them. This lets you see the negative densities as positive images. To get as accurate an interpretation of the densities as possible, use your knowledge of how these densities should appear on printing paper.

You know that a correctly exposed zone I density appears on the print as a maximum black, the same print tone as the density of the clear film base. If you expose your contact sheet with enough light to print the film base as maximum black, then all correctly exposed densities on each negative should appear in the zones in which you previsualized them. A correctly exposed zone III should appear as zone III, a correctly exposed zone VII as zone VII, and so forth.

For 35mm film, the easiest way to find the correct contact print exposure is to expose until the sprocket holes on the film disappear into the same print tone as the film base. Making a contact sheet any other way does not give you this information. This is why it is sometimes called a *proper proof*. The following illustration shows an exposure test for a proper proof:

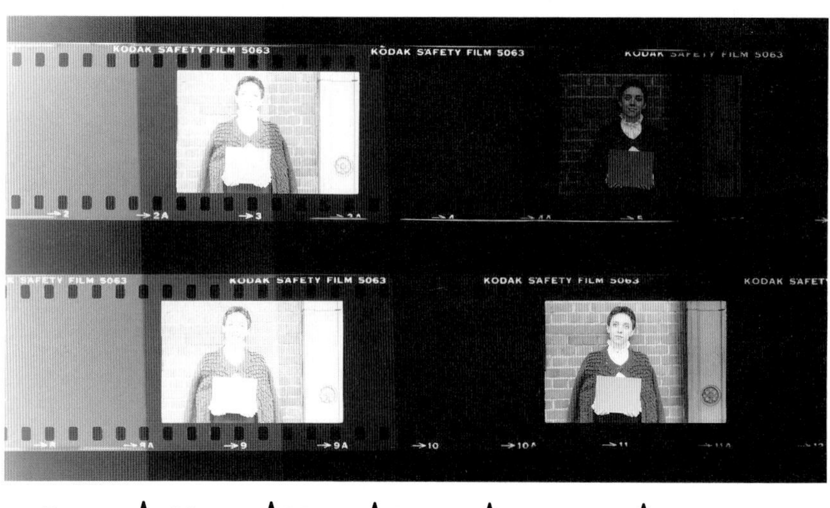

5 sec. ↑ **10 sec.** ↑ **15 sec.** ↑ **20 sec.** ↑ **25 sec. Maximum Black** ↑ **30 sec.**

A correctly made contact sheet gives information about exposure in addition to showing what the negatives look like. Make a proper proof by exposing the film base for a zone I value. The sprocket holes of 35mm film will literally disappear as the exposure reaches the point that the film base prints as maximum black. In this example, 25 seconds is the correct exposure time. Before that, the sprocket holes are clearly visible. Any exposure beyond that is overexposure.

When you examine the finished contact sheet, look first for inconsistent exposures. As the exposure index changes from the highest to the lowest in the test sequence, the images should look progressively lighter on the contact sheet. If this sequence looks out of order, or two adjacent frames appear to have nearly the same density, review your notes. As long as the indicated meter readings and the exposure placements are all correct, the problem may be caused by an inconsistent shutter. See Appendix B for a simple test that you can run if you suspect shutter problems.

If you do not detect a shutter problem, then the correctly exposed frames on the contact sheet will look approximately the way you previsualized the scene. Any underexposed frames will appear too dark, and any overexposed frames will appear too light. What you see in your contact sheet should confirm what you observed when you examined the negatives. Page 71 shows a proper proof made from the exposure index test negatives.

In this proper proof, the underexposed frames (numbers 3 and 5) appear too dark. The overexposed frames (numbers 11, 13, and 15) are too light. Frames 1, 7, and 9 seem to be the best. These are also the frames that had the correct amount of shadow detail in the negatives. This is another indication of the correct exposure index, somewhere between 200 and 400.

Printing the Exposure Index Test

The final and most accurate way to evaluate the results of your E.I. test is to enlarge the negatives. This is how you see negative densities as they actually appear in your prints. Although your choice of paper brand and paper developer typically does not have a great effect on the exposure index, keep in mind that the final print is an accumulation of all the choices you make, starting with your previsualization and ending in the darkroom.*

Begin with frame 1 (E.I. the same as the film's ISO), and make a standard size enlargement on normal contrast (grade 2) paper in which you try as carefully as possible to match the skin tones of the print to those of your subject. Most people tend to be very aware of how skin tones (especially the face) appear in black and white. If you print the skin tone a little dark or a little light, it is usually easy to notice it. Typical Caucasian skin is normally a zone VI; darker skin might appear as low as zone V. Regardless, this becomes a standard value for you to match as you print. If you are not sure how closely your prints match the actual subject, be sure to have the subject available for consultation.

* Although darkroom techniques are beyond the scope of this book, you should pay as much attention to your procedures in the darkroom as you do to your procedures for exposing film. A companion volume to this book, *The Elements of Black-and-White Printing*, teaches darkroom technique in a manner that is especially helpful to photographers using the zone system.

E.I. 1600: Shadow values in the hair (zone III) are too gray and lack detail.

E.I. 800: While the hair is darker than in the E.I. 1600 print, it still lacks detail.

E.I. 100: Satisfactory shadow detail; satisfactory highlights; noticeable grain in skin and gray card.

E.I. 50: Satisfactory shadow detail; highlight of zone VII blouse is too gray.

E.I. 400: Satisfactory shadow detail (in the hair); satisfactory highlights (the zone VII blouse); least amount of grain in the face and gray card; best overall appearance of tones.

E.I. 200: Satisfactory shadow detail; satisfactory highlights; some appearance of grain in the face and gray card; not quite as much contrast as the E.I. 400 print.

E.I. 25: Satisfactory shadow detail; highlight of zone VII blouse is too gray.

In this example of exposure index prints, the choice of the best overall exposure index was between the E.I. 200 and the E.I. 400 prints. It was the personal choice of the photographer to pick the E.I. 400 print. Another choice that is possible, if neither print had been significantly better than the other, would be to decide on an E.I. of 300. Halftone reproduction makes it impossible to show all of the detail present in the original prints. This might make it difficult for you to see the subtle differences mentioned in the text.

Once you make the first print and confirm that the skin tone matches your subject as closely as possible, print all the other negatives in the E.I. test to match the face in the first print. You will have to adjust the exposure of each print to do this. The more density a negative has, the longer the print exposure will be. You can skip frame 7 because this should be a duplicate of frame 1. Be sure to make all prints on normal contrast paper and always print the full negative without cropping.

CHOOSING YOUR EXPOSURE INDEX

When the prints are dry, you are ready to make the final determination of your exposure index. The basic principle of this evaluation is that the print made from the negative with the best exposure will look better than the other prints, in some significant ways.

Spread the prints out in a well-lit space where you can see them all at once. Avoid daylight-only or fluorescent-only lighting. Both types of light are rich in the blue part of the visible spectrum and tend to make it harder to accurately judge shadow detail.

Place the prints in order from the highest E.I. (4 times the film's ISO) on the left to the lowest E.I. ($^1/_{16}$ of the film's ISO) on the right. Then, check to see that the faces in all the prints match the actual skin tone and also each other. If they are slightly off this standard, you can make mental allowances when examining the prints. If they are more than a little off, either with respect to each other or to the actual skin tone, you should reprint the ones that are incorrect. A little extra time spent here will prevent you having to rely on guesswork later on.

Evaluating the Shadows

Begin your examination with the previsualized zone III object. Starting at the left, eliminate from further consideration any prints that do not show enough detail in this shadow. You should already have an expectation of which prints these will be, based on your examination of the negatives, but try to be objective. If, however, you find yourself keeping prints made from negatives that showed no detail, or eliminating prints made from negatives that showed definite detail, then stop and review, and do not proceed until you can discover a reason for this. If you do find a mistake, do not continue until you have corrected it.

Evaluating the Highlights

Next, examine the highlights. Starting with the prints on the right (with the lowest E.I.s), eliminate any prints that are too dark or gray, especially the zone VII object that you included in the image. Keep only those prints that you feel have acceptable highlights.

Evaluating the Midtones

By this time you should have narrowed the original seven prints down to only two or three. If not, repeat the first two steps. Each of the remaining prints should have good shadow and highlight tones that are close to, if not exactly, how you previsualized them. Each of these prints might well be acceptable prints by most standards. Yet, if you look closely, one print is better than the others.

Carefully inspect the grain of the remaining prints as well as the overall image quality. In looking for the best grain structure, look closely at the face and the gray card. The least mottled skin and the smoothest gray card indicate the finest grain. If these areas do not help, look at any relatively smooth area in zones IV to VI.

If, after eliminating any prints that show obviously inferior grain structure, more than one print is left, look at the overall image. Keep in mind the qualities of the tones that you have purposely included in the image. Try to see if one print shows all of these tones in the best possible way. Do not be misled by a good rendering of just one zone, look for the print that is best in all areas combined.

When you have one print left, the exposure index you tested for that negative is your new exposure index. If, as sometimes happens, you cannot decide between two adjacent prints, then choose an exposure index that is halfway between the two remaining prints.

Final Considerations

The exposure index that you find using this test may be the same as the manufacturer's ISO rating for the film, or it may be different. Experience shows that approximately half the exposure indexes found through testing are the same as the film's ISO rating. Whatever the case, your exposure index is now personalized to your equipment and allows you to know that when you place a shadow it will actually be in the zone you previsualize.

This exposure index test works because a film's response to exposure is exactly as the zone system predicts. An underexposed zone III will lack shadow detail in both the negative and the print. In the same manner, an overexposed negative will have highlights that are pushed against the negative's maximum density. Highlights on a print made from an overexposed negative will appear flat and darker than you previsualize.

Grain is also affected by exposure. The greater the density of a negative, the more grains of silver that are clumped together, and the more you will notice that grain clumping in the print. The optimum negative produced in this test is one that has enough density to show detail in a placed zone III shadow, but not so much density that it causes excess grain or gray highlights. Even though you can make a printable negative with a wide variety of exposures (as you proved with this test), only one exposure produces the best possible print when you look at it closely.

ALTERNATIVE EXPOSURE INDEX TESTS

The best tests in photography are tests that duplicate actual working conditions, such as the exposure index test described in this chapter. Other methods of finding an exposure index, however, are worth mentioning.

Densitometer Test

If you have access to a densitometer (a device that measures film density — many custom labs and university photography departments have them), then readings made right on the film will give you the information you need to calculate an exposure index. A densitometer measures the density of film based on a logarithmic scale from 0 (least density) to 3.0 (maximum density).

Procedure

Set your camera on a tripod in front of a gray card in even, normal contrast light (an overcast day or the north side of a building on a sunny day). Make sure that the image of the card fills the viewfinder and then throw the lens out of focus. Expose a roll of film in the following manner:

1. Set the light meter at an E.I. equal to the film's ISO rating. Take the meter reading from the gray card and place it in zone I (stop down four stops). Then expose the first frame.
2. Cover the lens and make frame 2 a blank.
3. Reset the light meter to 4 times the film's ISO rating, re-meter, and again place the indicated meter reading in zone I. Expose frame 3.
4. Cover the lens and make frame 4 a blank.
5. For the remainder of the test, use exposure indexes of twice the film's ISO, the same as the film's ISO, 1/2 the film's ISO, 1/4 the film's ISO, 1/8 the film's ISO, and 1/16 the film's ISO (the same as described in the empirical E.I. test). Remember to place the exposure of each frame in zone I.

Evaluating the Test

After you develop and dry the film, take densitometer readings from the film base (blank frames) and the zone I frames. The correct exposure index corresponds to the zone I frame that gives a density of between .03 and .06 greater than the film base reading. If you are testing 120 roll film or sheet film, the density difference you are looking for should be .10 to .15 greater than the film base.

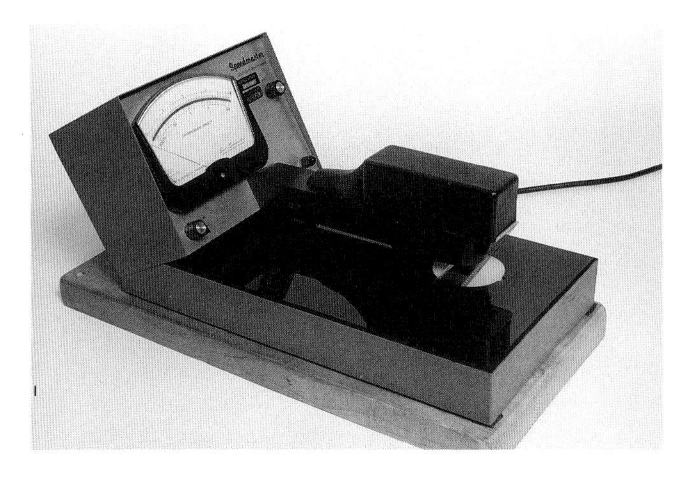

A densitometer can provide the information to calculate an exposure index directly from the film. Density information from this machine, however, cannot tell you what an image will actually look like. Older model densitometers, such as the one pictured, can often be purchased relatively inexpensively.

Print Tone Test

If you don't have access to a densitometer, an alternative to reading the densities is to compare how the tones look on printing paper. For this test, shoot a roll of film in a similar manner (using the same

E.I.s) as the above test, only placing the gray card in zone II instead of zone I.

Printing the Test Negatives

Print the processed film in the following manner. This test assumes that you are using fresh developer at the correct temperature and fresh enlarging paper to make all prints.

1. Starting with the highest E.I., select the first visible frame (some of the underexposed frames will have no visible density).
2. Place the frame you select in the negative carrier of your enlarger so that half of the adjacent blank frame is in the opening along with the zone II frame.
3. Focus the enlarger to make an approximately 8 × 10 inch print and stop the enlarger lens down to around f/8.
4. Expose a strip of grade 2 enlarging paper at increasing 5-second intervals, covering a small section at each interval.
5. Develop the print using your standard print processing procedures.

Evaluating the Test Print

When the print is dry, examine the effect of increasing exposure on the side of the paper exposed to the film base. Look for the step at which the film base prints as maximum black. At this point the enlarging paper will not get any darker, even as you increase the exposure. If this does not happen, even at the longest exposure time, make another print, either increasing the exposure time or opening the enlarging lens one stop.

At the point where the film base density prints as maximum black, look at the zone II side of the print at the same exposure time. It should show a definite lighter tone, just as zone II is a definite lighter tone than maximum black. If the zone II side does show this difference at a normal viewing distance and under normal room lights, then this frame represents the correct exposure index. If the zone II side does not, it indicates that even though visible on the negative, there is not enough density to print as a separate tone the way a zone II should. Re-run the printing test using the next darkest frame on the film.

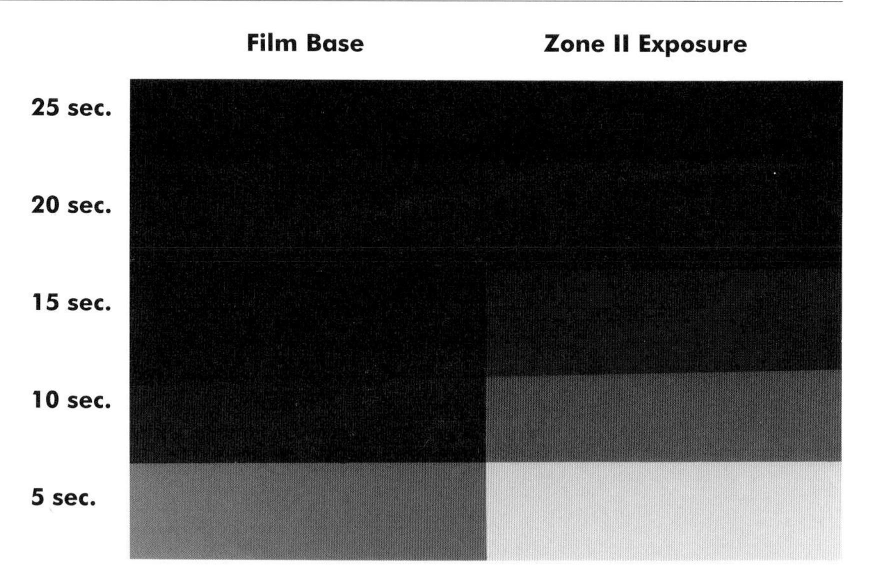

At 15 seconds the film base side of the test negative (see the text) has become maximum black. The zone II density at 15 seconds is noticeably lighter. This fits the zone system definition of the difference between zone I (maximum black) and zone II (first appearance of gray). This indicates the correct exposure index.

SUMMARY

Before you can use the zone system accurately, you must calibrate your equipment, materials, and working procedures to produce the film densities that you previsualize.

- An exposure index, or E.I., is the calibration that you use to compensate for possible errors in your camera shutter or light meter, and to take into account your particular film and developer combination.
- Use the exposure index number you find through testing to set the ISO dial on your light meter.
- The best test for exposure index is an empirical test that duplicates as much as possible the way you actually photograph.
- When you examine your exposure index test negatives, look for the first frame (going from least exposure to most exposure) that shows detail in the placed zone III area. This frame is a preliminary indication of your correct exposure index.
- When you examine a proper proof of your exposure index test negatives, look for the frames that print with the correct tones, not too dark and not too light. These frames should confirm the exposure index that you saw when examining the negatives.
- The final proof of the correct exposure index is to compare printed enlargements of your test negatives. Once you match the skin tones in all the prints, look for the print that has the best shadow detail, the best contrast in the highlights, and the least noticeable grain.
- If you cannot decide between prints that were made from two adjacent frames, use an exposure index that is halfway between the two.
- You can also use a densitometer, or you can compare tones that you make on printing paper to determine your exposure index, but these tests cannot tell you how actual images will look when you print them.

Chapter **9**

Finding Your Development Times

The photographer previsualized and placed the grass in this image in zone III rather than in zone V, which is the normal rendering of sunlit grass. A longer-than-normal development gave the grave stones their luminous quality. Knowing how to develop film for a specific contrast makes a nonstandard previsualization no more difficult than a conventional way of seeing the world.

Once you find the correct exposure index for your camera, you are ready to determine your normal contrast (N), low contrast (N+), and high contrast (N−) development times. This means finding out how long it takes to develop your film so that the highlight densities match your previsualization.

DEVELOPING FOR THE HIGHLIGHTS

The film development graph on page 45, illustrates how development time affects a negative's contrast. The longer film develops, the greater the contrast. The trick is to stop development at precisely the moment that the negative densities match your previsualization of the zones in the scene.

In theory, every film and developer combination should have the same normal, high, and low contrast development times, regardless of who uses that combination. Unfortunately, the effect that agitation techniques, water quality, and even the type of developing tank that you use can make your normal development time drastically different from another person's. There is no recipe or mathematical formula that can take into account all the variations in individual processing methods.

This means that you cannot get your development times from a book or another person. You must find them for yourself. You must test each development time you plan to use. The best that someone else's data or a manufacturer's recommendations can give you is a starting point for your tests.

Fortunately, testing is straightforward once you apply the zone system to the problem. The basic steps are as follows:

- First, previsualize a scene so you know how you want the highlights and shadows to look on the print.
- Next, use exposure placement to determine the shadow densities on the negative. The shadows will not change significantly during development.
- Finally, test development times until you find the time that produces the important highlight densities the way you previsualized them.

Testing in this manner is not difficult, but finding the correct highlight density in this way is time consuming. Plan to set aside a few hours each day for the next several days to complete your development tests.

Effect of Enlarger Type

The type of enlarger that you use affects the development time of your negatives. There are two basic types of enlargers in common use, *condenser* and *diffusion*.

A condenser enlarger uses large glass lenses above the negative to concentrate and focus the light before it passes through the negative. A diffusion enlarger uses a piece of frosted glass or plastic just above the negative to scatter the light nondirectionally. A variation of the diffusion enlarger uses a special fluorescent tube as a light source and is called a *cold-light* enlarger. Each enlarger type requires a different contrast negative to print properly.

The focused light of a condenser enlarger produces something called the *Callier effect** when that light passes through the negative. This effect is characterized by a scattering of light that is more pronounced in areas of greater density than areas of lesser density on the negative. As a result, a print you produce with a condenser enlarger has greater contrast than a print you produce with a diffusion enlarger.

The implications this has for development time is that a negative you intend to print with a condenser enlarger requires less development than a negative you intend to print with a diffusion enlarger. This is true for any film and developer combination, and for any contrast scene.

In a more subtle way, every make of enlarger, even those of the same type, has its own unique effect on contrast. The same is true of enlarging lenses. The entire process of exposing, developing, and

* Hollis N. Todd and Richard D. Zakia, *Photographic Sensitometry* (Dobbs Ferry, NY: Morgan & Morgan, 1969, p. 193–194)

printing photographic images is a chain in which each link leaves a unique mark. As you work through the tests in this book, try to remember that you are testing all the links in your system. If you change one of the links in the future, be aware that it might require new testing, or at least modifying one or more of your procedures.

The image on the left was printed with an Omega condenser enlarger. The image on the right was printed with the same enlarger modified by adding a "cold light" diffusion light source manufactured by Aristo Grid Lamp Products (Port Washington, NY). The examples shown are made from a negative that was developed for a condenser enlarger. As you can see, what is correct for the condenser enlarger is too flat for the diffusion model.

NORMAL CONTRAST DEVELOPMENT

To find your developing time for a normal contrast scene, use the following steps:

1. Find a scene with normal contrast. Carefully check the contrast range between the previsualized shadow zones and highlight zones. The scene should have large, easily recognizable shadow areas, especially zone III, and an equally large area of highlights, especially zones VII and VIII.
2. Place the exposure for the most important previsualized shadow.
3. Expose an entire roll of film. A tripod will help make each frame as nearly identical as possible, but it is not essential unless your exposure times are slower than $^1/_{60}$ of a second.
4. In the darkroom, cut the film into approximately three equal lengths. Save two lengths in a light-tight container, and develop the third for the manufacturer's recommended development time, or whatever you consider to be a normal development time for that film and developer combination.
5. When the film is dry, test the contrast by printing one of the frames on normal contrast paper so that the important highlight appears in the zone in which you previsualized it.
6. Once the print is dry, confirm that the highlight is in the correct zone and then examine the placed shadow as follows:
   - If the shadow appears the way you previsualized it, then the development time is correct for a normal contrast scene.
   - If the shadow appears too dark — for example, if a previsualized zone III area lacks detail — then the negative is too high in contrast and overdeveloped. See the section *Reducing Test Time* on page 83 for instructions on what to do next.

- If the shadow appears too light and gray — for example, if a previsualized zone II area appears gray and has detail — then the negative is too low in contrast and underdeveloped. See the section *Increasing Test Time* on page 83 for instructions on what to do next.

When the tones on a print made on normal contrast (grade 2) printing paper match your previsualization, that indicates the film development is correct. The drawing shows the previsualization for a normal contrast scene. The images illustrate three different development times for the negative. The correctly developed negative renders both the zone III and zone VII areas as previsualized. The overdeveloped negative renders the zone VII highlight correctly, but the zone III area appears too dark and lacks adequate detail. The underdeveloped negative renders the zone VII highlight correctly, but the zone III tone is too light and gray.

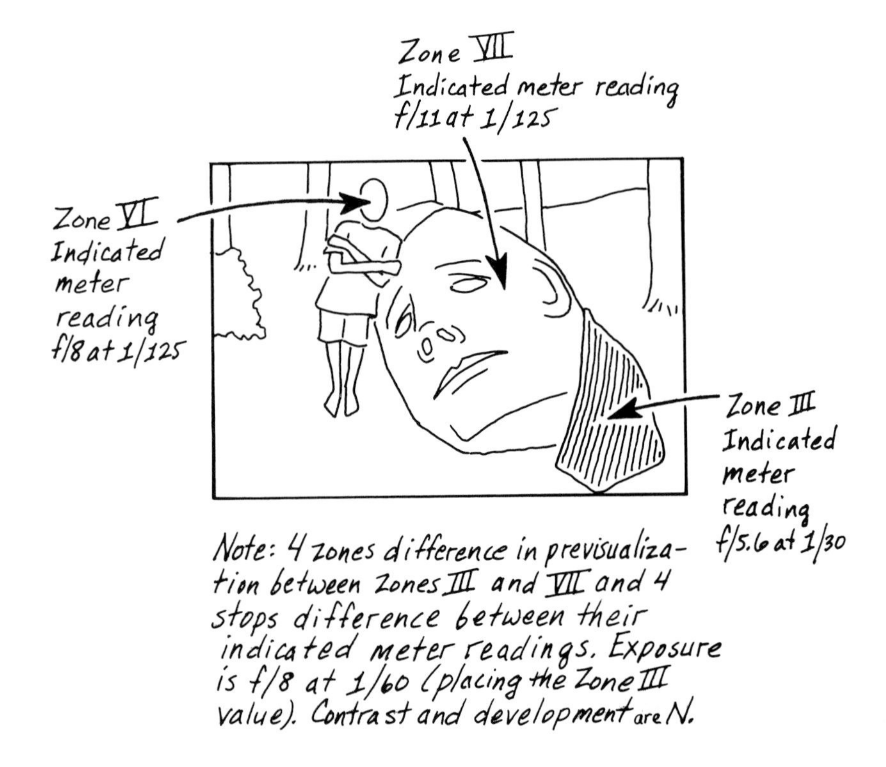

Correctly Developed Negative

Overdeveloped Negative

Underdeveloped Negative

Why Evaluate Print Shadows?

It might seem confusing that you base your film development time on how the shadows in the print look, especially because development primarily affects the film's highlights. There is, however, a logical reason for this. What you are evaluating when you make a print of the test negative is the overall negative contrast.

You can best evaluate contrast in a print when the exposure is correct for the highlights. The highlight is the area of least density in the print and, like the shadows in the negative, is the area most sensitive to changes in exposure. The highlights, therefore, are how you determine correct print exposure.

Only in a correctly exposed print can you judge its overall contrast. Even though you might see detail in the zone III area of an overdeveloped negative, that detail will be suppressed in a correctly exposed print made on normal contrast paper. The effect on a print you make from an underdeveloped negative is similar. The shadows in that print will appear lighter than you previsualized them.

Reducing Test Time

If you determine that the first segment of film is overdeveloped, then take the second segment and develop it for less time. How much less depends on your perception of how much extra contrast appeared in the print.

- If the zone IV areas of the print looked like a zone III and the zone III areas had no detail at all, then try developing about 7% less for low-speed film (ISO 25 to 50), 10% less for medium speed film (ISO 100 to 200), and 15% to 20% less for high-speed film (ISO 400 or higher). A rule of thumb is that the lower the light sensitivity of a film emulsion, the more rapidly it reacts to changes in development.
- If the zone IV areas of the print were too dark to even have detail, then try doubling the percentages.
- If the print was just a little too contrasty, such as the zone III having a slight but not really sufficient amount of detail, then try reducing development by less than the above percentages.

After testing the second segment of film, you have the third segment available if you need to make further adjustments in your development time, or as a backup if something goes wrong in processing one of the other two.

Increasing Test Time

If you determine that the first segment of film is underdeveloped, then increase the development of the second segment by the same percentages that you would have decreased it for too much contrast. Again, look at the way the shadow values appear in the print. If the shadows appear lighter than your previsualization, then you need to try a longer development time. If the shadows appear darker than your previsualization, then you need to try a shorter development time. If you are careful and use the same temperature and agitation method each time, you should be very close to finding the correct development time by the third segment, if not the second.

LOW CONTRAST (N+) AND HIGH CONTRAST (N−) DEVELOPMENT

The test for finding development times for N+ and N− contrast scenes is similar to the test for normal contrast. Modify the procedures for finding your normal contrast development time in the following way:

1. Find a scene with the type of contrast you plan to test. Make sure that the scene has large areas of easily identifiable zones II and III shadows, and zones VII and VIII highlights.
2. Shoot an entire roll of the scene, making sure your exposure placement is as accurate as possible.

3. Cut the film into thirds, put two of the segments into a light-tight container, and develop the third in the following manner:
   - For N+1 contrast, begin with a development time that is longer than your normal contrast time by approximately 20% for high-speed film, 10% for medium speed film, and 7% for slow-speed film.
   - For N−1 contrast, reduce your normal contrast development time by the same percentages (20% for high-speed film, 10% for medium speed film, and 7% for slow-speed film).
4. Print one of the frames from the processed film on normal contrast paper so that the important highlight appears in the zone in which you previsualized it.
5. When the print is dry, examine the placed shadow to see if it fits the correct description for the zone in which you placed it.
   - If the shadow is too dark, then decrease the development time for the next segment of film by approximately 20% for high-speed film, 10% for medium speed film, and 7% for slow-speed film.
   - If the shadow is too light and gray, then increase the development for the next segment of film by approximately 20% for high-speed film, 10% for medium speed film, and 7% for slow-speed film.

If you are careful, you can complete these tests by the second or third segment of film. With 35mm film, you rarely need more than N, N+1, and N−1 development times. If you ever need other development times, you can test for them later.

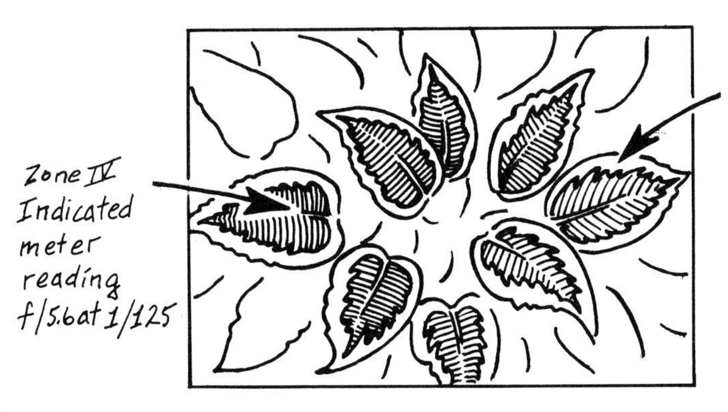

In a low contrast scene such as this, the film needs additional development (N+1) to increase the separation between the highlights and shadows. The sketch shows the previsualization and indicated meter readings for the scene. At the normal (N) development time, the zone IV appears on the print to be more like a zone V. With an N+1 development time, the negative renders the zone IV as it was previsualized.

Normal Development

N+1 Development

CHANGING YOUR CALIBRATIONS

Like the exposure index test, the film development test is empirical. The results depend on how you actually make and evaluate your own prints. Your development times, therefore, are based on your interpretation of print tones at the time you did the testing. Remember that each zone is a segment of the continuous tone scale and can appear many different ways in the final print. It is true that these are subtle differences, but as your awareness of and sensitivity to print tones grow, how you want your prints to look may well change. Expect either your exposure index or development times to change in response to your growth as a photographer.

One way to monitor this change is to keep notes of the paper grade on which you consistently print. The ideal for zone system photographers is to print on grade 2 paper, because the normal contrast grade renders tones most closely to the way the densities appear on the negative. If you find that your final prints are consistently on a paper grade other than grade 2, then you should consider changing your development times. For example, if you are constantly printing on grade 3 paper, then your development time should be longer. If you are constantly printing on grade 1 paper, then your development time should be shorter.

In a high contrast scene, the film needs a lower-than-normal development time (N−1) to decrease the separation between the highlights and shadows. The sketch shows the previsualization and indicated meter readings for the scene. At the normal (N) development time, the zone IV appears on the print to be more like a zone III. With an N−1 development, the negative renders the zone IV as it was previsualized.

Normal Development

N−1 Development

Effect of Development on Exposure Index

Changing development times for contrast control has a small but real effect on the shadow densities of your negatives. You can see this effect in the graph on page 45, where the curve for the shadow densities is not a completely flat line.

The need to change your exposure index is most noticeable when development exceeds N+1 or N−1. If you occasionally use N+2 or N−2 development times, you can generally reduce your exposure by about one-half stop for a negative that you are going to develop for N+2, or add about one-half stop of exposure for a negative that you are going to develop for N−2. If you consistently use one of these development times, then you should retest your exposure index for that contrast. Note that most 35mm film is not well suited for development times that exceed N+1 or N−1.

ALTERNATIVES TO THE DEVELOPMENT TEST

Just as for the exposure index test, there are alternatives to the empirical development test. The following sections describe ways you can find your development times by using a densitometer or by matching the tones you produce with your enlarger.

Densitometer Test

With a densitometer, the development test consists of taking density readings from a zone VIII tone. Zone VIII is the best value to use in a densitometer test because it is the highest negative density that has a visible tone on printing paper. This makes it the tone most visibly affected by development changes.

What constitutes a proper zone VIII density depends on the enlarger type. Diffusion and cold-light enlargers require a lower contrast negative than a condenser enlarger. The best way to find the correct highlight density for your equipment is to take densitometer readings of the zone VIII value on at least 10 of your negatives that print well. Throw out any unusually high or low readings, and average the rest. This is the zone VIII density that works for your system. The following table lists starting points for comparison:

Film Type	Density for Condenser Enlarger	Density for Diffusion Enlarger
35mm	1.15	1.25
120 roll or sheet film	1.30	1.40

Normal Development

Use the following procedure to find a normal development time:

1 Shoot a gray card (out of focus and filling the viewfinder) placed in zone VIII.

2 Cut the film into thirds and develop the first third for what you believe is the normal development time.
   - If the density reading is higher than the target zone VIII density, develop the next segment for less time.
   - If the reading is lower than the target zone VIII density, develop the next segment longer.

A difference of .30 from the target density suggests that you should change your next development time by 7% for low speed film, 10% for medium speed film, and 15% to 20% for high speed film.

N+1 and N−1 Development

Testing for N+1 development involves the same procedure as the test for normal development, except that the zone you place should be zone VII. Look for a development time that develops the placed zone VII to the target zone VIII density.

The test for N−1 development requires that you place the gray card on zone IX. The development time you are looking for is the time that develops this zone IX exposure so that it matches the target density for zone VIII.

SUMMARY

Once you find your exposure index, the next step is to find correct development times for normal (N), low contrast (N+1), and high contrast (N−1) scenes.

- The areas of greatest density (highlights) on the film are the values that you use to determine development time.
- The type of enlarger that you use affects how long you develop film. Condenser enlargers generally require lower contrast negatives than diffusion or cold-light enlargers.
- You can test for normal contrast development time by photographing a normal contrast scene and printing the negative on normal contrast paper. When the print contains both the highlight and the shadow zones you previsualized, then you have confirmed the correct development time.
- When evaluating prints for the correct film contrast, start by making sure that the print exposure renders the previsualized highlight in the correct zone. Then, examine the previsualized shadow to see if it appears in the correct zone on the print.
- When reducing or increasing development times during your test, a rule of thumb is that the slower the film emulsion speed, the faster it responds to changes in development.
- You can test for N+1 development time by photographing a low contrast scene and printing the negative on normal contrast paper. When the print contains both the highlight and the shadow zones you previsualized, you have confirmed an N+1 development time.
- You can test for N−1 development time by photographing a high contrast scene and printing the negative on normal contrast paper. When the print contains both the highlight and the shadow zones you previsualized, you have confirmed an N−1 development time.
- The development test, like the exposure index test, is based on how you see your previsualized zones on a print. As you become more fluent in the language of the zone system, your visual sense will change and may require that you change some of your calibrations.
- You can use a densitometer or match enlarger tones to find your film development times. These tests offer the advantage of saving time at the expense of losing a sense of how the tones print on your printing paper.

Chapter 10

Shortcuts for 35mm Cameras

In some scenes, it is nearly impossible to walk up and meter specific tones. If you have a previsualized substitute nearby and it is in the same light as the scene you are photographing, then you can still use the zone system to find the exposure and contrast of the scene.

If the zone system truly offers a way to grasp the creative potential of photography, then you should be able to adapt previsualization, metering, and contrast determination to virtually any camera or film format. You should not have to change the type of camera you use or the format of your film to enjoy the technical control of the zone system.

ADAPTING THE ZONE SYSTEM

To some photographers using the zone system, speed is not important. A photographer with a large format camera on a tripod, for example, is usually photographing subjects that are not moving. In such a case, there is time to previsualize all the important tones in the scene and take careful readings of the light reflected from each one. Handheld spot meters make the process of taking these readings an easy, if deliberate, matter.

Photographers using 35mm cameras, however, have the ability to move with rapidly changing scenes, an ability that comes from the portability of a small, self-contained unit. When the action is fast and furious, it is tempting to give up the control that the zone system offers for convenience. This is not necessary. You can adapt the zone system to a handheld camera with a built-in meter and still photograph action.

Previsualization Substitutes

When you cannot take light meter readings of distant or very small objects, you can previsualize and meter tones that are close and easily accessible, and then relate them to the scene you want to photograph. This requires tones that must appear in a specific way in a print to avoid the risk of altering the viewer's perception that the tones are rendered naturally. Such a tone is the zone VI previsualization of Caucasian skin. Most viewers have an exact idea of what this tone looks like in a print and can sense whether or not it is rendered correctly, even if they are not photographers.

When this skin tone appears partially lit and partially in shadow, only the most brightly lit portion remains a zone VI to the eye. The shaded portion becomes a zone IV. This is the standard previsualization of a light skin tone in a photograph. The skin must appear in these zones if you want the tones in the print to appear the way the eye sees them in the original scene.* Most other objects can be lighter or darker by as much as a zone and a viewer will not notice the difference as easily.

The face of the subject in this portrait is in uneven light. The most brightly lit side is normally a previsualized zone VI, while the shaded side is a previsualized zone IV.

The Palm of Your Hand

The standard previsualization of Caucasian skin tone would be of limited use except for the fact that virtually everyone has an equivalent tone, in terms of light reflectance, right in the palm of

* This is true for all light-skinned people. Although the majority of dark skin is previsualized as zone V, there is too much variation to make a general rule. You can establish a personal standard for dark skin tone by comparing it with a known value such as a gray card.

their hand. The amount of light reflected from the palm is generally the same for most people, regardless of skin color. In the brightest ambient light, the palm is a previsualized zone VI; when shaded in the same light, it is a previsualized zone IV.

You should confirm this with the following experiment:

1. Find a location with even light, such as the north side of a building on a sunny day.
2. Hold up your palm and move it around (with your eyes squinting if necessary) until it is brighter than at any other point. This should be when it is pointing directly at the main source of light. If you are standing on the north side of a building, the light source will be the north sky.
3. Take a meter reading from your palm and write it down.
4. Place a gray card in the same light, take a meter reading, and again, write it down.
5. Next, move your hand around until the palm appears darker than at any other point. This will occur when the back of your hand is blocking the main light source. Your palm should be turned 180 degrees from where it was when you made the first meter reading.
6. Take a meter reading from your palm in shadow and record it.

If the indicated meter reading of your hand in light indicates one stop more exposure than the reading of the gray card, then you have confirmed a zone VI previsualization for your palm. If the two indicated meter readings of your hand (in light and in shadow) are two stops apart, then this indicates the difference between zone IV and zone VI in a normal contrast scene.

If you do not have a two-stop difference, then your readings were in uneven light (either too directional or too high in contrast), or in light that does not have enough intensity. Try another scene. North light between 10:00 A.M. and 3:00 P.M. on a sunny day will always produce a two-stop difference.

FINDING EXPOSURE AND CONTRAST

With just two indicated meter readings, one from a previsualized shadow and one from a previsualized highlight, you know everything that you need to expose and develop your film.

Expose the film by placing the indicated meter reading from your shaded hand in zone IV (close down one stop). Determine the contrast by comparing the difference in meter readings to the difference in previsualized zones.

A two-stop difference equal to the difference between zones IV and VI indicates a normal contrast development. A three-stop or more difference between meter readings (as will happen in direct sunlight) indicates high contrast. A three-stop difference is N−1, a four-stop difference N−2, etc. Conversely, a one-stop difference between meter readings indicates low contrast (N+1).

A Handy Shortcut

Rather than being limited to portraits, you can apply a zone VI highlight and zone IV shadow previsualization to any photograph.

Your hand does not have to be in the scene as long you make your meter readings in the same light.

This application of the zone system simplifies and standardizes the process of previsualizing and metering. Your hand is a flexible and easy to use target, and the palm is large enough to fill the viewfinder of a 35mm camera with a normal lens when you hold it at a comfortable distance. Remember that you do not have to have your hand in focus to make an accurate meter reading.

You can relate the light reflected from the palm of your hand to the tones in a scene in which your hand does not even appear, as long as you meter it in the same light as the scene. This is a valuable technique when you cannot walk up to meter subjects such as the solitary rower photographed from an overhead bridge.

Previsualization Test

Knowing that the palm of your hand can be either a zone IV shadow or a zone VI highlight also gives you a method for testing your previsualization. You can compare any suspected shadow zone with your shaded palm. If the indicated meter readings are the same, then you can previsualize the tone as a zone IV. If the shadow area reads one less stop than your hand, it is a zone III, and so on.

The same is true for highlights. You can compare the indicated meter reading of your palm in light with the indicated meter reading of a highlight you want to test. This test works in all but the most extreme high contrast and low contrast scenes, and is a good method to use when you are learning the zone equivalents for common objects such as those listed in Chapter 2, *The Zones*.

Benefits and Trade-Offs

Using the palm of your hand as a reference point for both highlight and shadow zones gives you flexibility and convenience. It does not matter whether a scene is next to you, across the street, or as far away as a distant landscape. Objects in the scene reflect light in the same relationship as the substitute tones that you have previsualized and measured.

You gain several benefits by using the zone system this way:

- You do not have to disturb a scene (such as a group of people) by walking up to meter specific tones.
- If a scene is too far away to approach, the substitute previsualized zones are close at hand.

- Once you select a location where you want to make photographs, your initial exposure and contrast determinations will not change unless the light intensity changes.

When you combine the technique of previsualizing and metering your hand with other shortcuts that are especially convenient with hand-held cameras, such as zone focusing or setting the lens to its hyperfocal distance,* the zone system actually simplifies your choices. You are free to concentrate on the image in the viewfinder rather than constantly making decisions about exposure and focus.

Of course, this method, like most shortcuts, is a compromise. It is always more accurate to previsualize and meter more than two zones, especially zones that are farther apart than zones IV and VI. For most situations, however, this method is quite reliable.

The primary limitation of this technique occurs when you want to use a nonstandard previsualization for a scene. The zone VI hand-in-light, zone IV hand-in-shadow previsualization produces negative densities that match how the eye normally sees tones in a scene. If you want to place the shadows higher or lower, or develop the film for lighter or darker highlights than this standard, then you must measure the tones in the scene directly.

MAKING THE SHORTCUT WORK

Putting the theory of this zone system shortcut into practice is a matter of understanding a few working procedures. First and foremost, you must be sure that you are standing in the same light as the scene you plan to photograph before you take any meter readings.

Any time you are outdoors on a sunlit or evenly overcast day, you can assume the light is the same for both you and your subject. This is also true if you and your subject are in the same shaded area. More often than not, you will be in the same light as the scene you are photographing.

Be cautious, however, when your subject is in the shade, such as under a tree, and you are standing in the sunlight (or vice versa). Also be aware that in scenes lit by artificial light, light intensity diminishes rapidly as the distance from the light source increases. You rarely find evenly lit scenes indoors.

The Highlight Reading

For the highlight reading, position the palm of your hand so it reflects the brightest light in the scene. A light source does not have to be the sun or an electric light — it can be an object that reflects a large percentage of light relative to other objects in the area. Be very careful to observe where your hand reflects the most light, especially in a situation where there is no direct light source.

---

* *Zone focusing* means using an estimate of the distance to the main subject for setting the lens focus and allowing the depth of field to make up for slight inaccuracies. *Hyperfocal distance* means setting the focus for the closest distance at which the depth of field also includes the infinity mark. On a bright day, hyperfocal distance might include everything from 10 feet to infinity. These techniques save you from having to constantly change focus when all of the subjects you are photographing are about the same distance from the lens.

Fill the viewfinder of the camera without casting a shadow on your hand (this may mean holding the camera at a slight angle) and take a meter reading. This is your indicated meter reading for a previsualized zone VI.

Take a zone VI reading from the palm of your hand at the point where the palm reflects the greatest amount of available light. The technique is simple: Hold your hand close enough to the camera to fill the viewfinder (it does not have to be in focus) and record the indicated meter reading. Be careful that the camera lens does not cast a shadow on your hand.

The Shadow Reading

For the shadow reading, rotate your hand 180 degrees so that it blocks the light source in the scene. This is a previsualized zone IV. Take a meter reading and stop down one stop to place the exposure in zone IV. If the light source is sunlight or strong artificial light, be careful that your hand completely shades the viewfinder; otherwise, the direct light hitting the meter cell may temporarily blind it.

Check the shadow reading from the palm of your hand often until you feel confident about your ability to sense exposure changes. The exposure will change, for example, when the sun goes behind a cloud, or when you walk into the shadow of a building or under a tree.

When your hold your hand between the light source and the camera, the palm becomes a previsualized zone IV. To make the reading, hold your hand close enough to fill the viewfinder. Be careful not to let direct sunlight strike the meter cell, or your meter could be temporarily "blinded."

Development

With roll film, you must choose your developing time carefully, as you can develop a roll only one way. Since contrast can change in the middle of a roll, consider the specific needs of each frame and then make a choice based on a few basic rules.

If you can, choose a development time that fits the majority of the frames on that roll. Contrast changes less often than exposure, so make a note when a change does occur and plan development to fit as many of your frames as possible. This is a good argument for shooting 20-exposure rolls instead of 36, or even bulk loading shorter lengths, so that you can shoot an entire roll before you move to a location with a different contrast. An alternative to this rule is to develop your film for the contrast of the most important frames on the roll.

All things being equal, on a roll of film with two different contrasts, you are better off developing for the higher contrast scenes (that is, developing for the least amount of time). You can add contrast in printing more easily than you can reduce it.

SPECIAL CONTRAST SITUATIONS

Even with basic ground rules, you may encounter situations where it is hard to know how to expose or develop a roll of film. This is especially likely when you encounter unusual or extreme contrast situations, such as scenes with very low contrast or an unusual mixture of contrasts on the same roll of film.

Very Low Contrast

You may encounter scenes in which there is almost no difference between the indicated meter readings of your hand in shadow and your hand in light. When the contrast is this low, use the indicated meter reading for your hand in light and place it in zone VI. Develop the film for normal contrast, if possible. The negative will print well on high contrast paper.

Different Contrasts on the Same Roll

A mixture of different contrast scenes on the same roll presents the greatest challenge. The following are a few suggestions to help you deal with the most difficult of these situations:

- If you must shoot an N+1 scene on a roll that you have already decided to develop for normal contrast, ignore the shadow reading. Instead, take the indicated meter reading for the hand in light and place it in zone VI. Print this negative on high contrast paper.
- If you must shoot an N−2 scene on a roll that you have already decided to develop for normal contrast, then ignore the shadow and place the indicated meter reading for the hand in light in zone VI. Print this negative on low contrast paper.
- Possibly the most difficult situation is an N+1 scene on a roll that you must develop for N−1 contrast. In this situation, place the indicated meter reading of the hand in light in zone VI ½ or VII (1½ or 2 stops more than the indicated meter reading). When you develop the film for N−1 contrast, the highlight density of that frame will be about zone VI. Print the negative on high contrast paper.

These are not ideal solutions, because sometimes you will lose highlight detail in the print. Remember that these are compromises to solve the problem of having different contrast situations on a roll of film that you can only develop one way.

USING PAPER GRADES

You can handle most of the difficulties of having different contrast scenes on a single roll of film through paper grades. When you place your exposures correctly, the shadow densities will be correct, even if the highlight densities vary from frame to frame. When the contrast of a negative is too high or low for the development you give a roll of film, you can use paper grades to correct it.

Usually, changing one paper grade equals a change of one zone in contrast. This means that you can print an N+1 scene on a roll developed for normal contrast with grade 3 paper or a number 3 filter.

The following table suggests paper or filter grades to use when the contrast of the scene does not match the film development:

Film Development	Contrast of Scene	Paper or Filter Grade
N	N+1	Grade 3
	N−1	Grade 1
	N+2	Grade 4
N+1	N	Grade 1
	N+2	Grade 3
	N−1	Cannot be printed
N−1	N	Grade 3
	N+1	Grade 4
	N−2	Grade 1

Although these guidelines can help you produce good results, keep in mind that you will get the best possible prints with a negative matched to a normal contrast paper grade. Paper grades other than grade 2 do not translate the tones in a negative as accurately as the normal contrast grade. Higher contrast papers pull tones farther apart as a corrective for low contrast negatives. Lower contrast papers push tones closer together so that high contrast negatives will look normal. This alteration of tones is not quite the same as changing contrast through film development.

The image of the dulcimers on top is an N+1 scene that was given normal development. It was printed on grade 3 paper to bring out the highlights on the instruments. The image of the wedding cake on the bottom is an N−1 scene that was given normal development. It was printed on grade 1 paper to keep the highlights on the white table cloth from becoming too bright. In both examples, the tones appear as previsualized, and required a minimum of burning and dodging.

SUMMARY

The strength of the zone system lies in its ability to adapt to the special needs and abilities of 35mm cameras.

- You can use the standard previsualization of the palm of your hand as a substitute for previsualizing tones in an actual scene, as long as your hand and the scene are in the same light.
- When the palm of your hand is pointing at the light source in a scene, it is reflecting the same amount of light as a previsualized zone VI.
- When the palm of your hand is blocking the light source in a scene, it is reflecting the same amount of light as a previsualized zone IV.
- While metering the palm of your hand, be careful not to hold your hand so close to the camera that it casts a shadow on your palm, or so far away that background light affects the reading.
- Compare the meter readings for your hand in light and hand in shadow to determine contrast. A difference of two stops indicates normal contrast, less than two indicates low contrast, and more than two indicates high contrast.
- When there are frames with different contrasts on the same roll of film, plan your film development to fit the contrast of the most important frames, or if all the frames are equally important, the majority of the frames.
- In a scene with very low contrast, place the indicated reading of your hand in light in zone VI and develop normally. Print the negative on high contrast paper.
- If you must shoot an N+1 or an N−2 scene on a roll that you have already planned for normal contrast development, place the hand in light reading in zone VI. Print the N+1 scene on high contrast paper and the N−2 scene on low contrast paper.
- If you must shoot an N+1 scene on a roll that you plan to develop for N−1 contrast, place the indicated reading of your hand in light in zone VI$\frac{1}{2}$ or VII. Print the negative on high contrast paper.
- If you cannot develop each frame on a roll of film for its optimum contrast, you can print on a paper grade that best matches the contrast of the negative.

Chapter 11

The Zone System and Color

At first appearance, color seems to add reality to a photographic image. In fact, the opposite is often true. Cover the bottom half of this image and you remove all the color from it. This does not make the image any less of a depiction of "reality" than when the whole image (and the color) is visible.

Using the zone system with color film is not much different than using the zone system with black-and-white film. You need the same ability to previsualize a scene and place the exposure. The major difference when using color film is that you must previsualize the relationship of colors in addition to the light reflectance of highlights and shadows.

ACCURACY OF COLOR

Many people have the impression that color film produces images that are a better depiction of reality than images produced with black-and-white film. In fact, the added dimension of color often creates a departure from an accurate view of the world, even allowing for the difficulty of defining "reality." The many reasons for this have to do with the nature of the color materials and with the limitations of our language for describing colors.

Film and Color Reproduction

The reproduction of colors in color film depends on the light response of multiple layers of silver emulsion and the subsequent replacement of the silver with color dyes. Each color emulsion renders colors differently depending on how the film was manufactured. Currently, some films, such as Kodak's Ektachrome 64 Professional and 100 Professional, tend to emphasize cool (blue) colors. Other films, such as Fujichrome Velvia, emphasize warm (red) colors.*

Any generalization about the nature of color films, however, is subject to change whenever the process for making them changes. In the past decade or two, manufacturers have increased the ability of their films to render deep, bright colors at the expense of color accuracy. This is an apparent effort to satisfy a consumer demand and is subject to change whenever there is a perception that the public taste in color photography has changed.

No matter what manufacturers do, however, color transparencies and prints can never reproduce completely accurate color. The dyes in the emulsion contain cannot duplicate the exact portion of the light spectrum that the original scene reflects. In every case, color materials interpret the colors in a scene.

Language and Color

An often repeated story about the relationship of language and color states that Eskimos have 17 or more words for the color *white,* each one defining a particular condition of snow. Although apparently untrue,† this story endures because of how it highlights the relatively limited color vocabulary of the English language.

In English, there are fewer than a dozen commonly used words for colors. Even the most traditional words that describe color (such as *pink* or *brown*) actually refer to areas of color that can contain many different shades, rather than a specific color.‡ This limits our ability to precisely describe all but a few color differences. Symptoms of this problem appear in the convoluted adjectives used by decorators and paint manufacturers for naming colors.

The lack of precision often forces people to make vague generalizations about what they see, not just with words, but in how they think about color. Josef Albers wrote that when you ask a roomful of people to visualize the color *red,* everyone will think of different shades of that color. Even when you specify a familiar shade of red, such as the red used in Coca-Cola advertisements, the results will be the same. You can even show various shades of red to an audience and ask individuals to pick out the shade that matches their memory of the read that appears on a "Coca-Cola" sign, and there will be disagreement.

Albers thought that this was due to the fact that people have a poor memory for color. Whereas most people can remember sound

* Marc Levey, *The 35mm Film Source Book* (Boston: Focal Press, 1992, p. 92).

† Geoffrey Pullum, "The Great Eskimo Vocabulary Hoax," in *Lingua Franca* (June 1990).

‡ A.H. Munsell, *A Color Notation* (Baltimore: Munsell Color Company, Inc., 1961, p. 52).

patterns (such as the beat of a popular song) for long periods of time, their memory of color is fleeting.*

Most people have a poor memory for color. Look at the shades of red here and try to decide which is closest to matching a familiar red color, such as that used in "Coca-Cola" advertising. Make a guess and then compare. Your guess will probably be different from the actual color.

DESCRIBING COLOR

Albert H. Munsell identified three distinct qualities that every color possesses: *hue, value,* and *chroma.*†

- *Hue* is the name of a color, such as green or red. It refers to a specific wavelength or part of the color spectrum.
- *Chroma* (frequently called *saturation* by photographers) indicates the strength or intensity of a hue. A hue mixed with any other hue, or black or white, will change the chroma, or intensity, of that hue. A pastel color, which has a hue mixed with a lot of white, has a weak chroma (color photographers usually say it is *desaturated*). Vivid hues have a strong chroma, or are *saturated.*
- *Value* (also called *brightness* or *luminance*) refers to the lightness or darkness of a color as it appears next to other colors in a scene.

These three terms can describe any color in any situation. For example, red is a specific hue, the longest wavelength of visible light. Pink is a red with a weak (desaturated) chroma, one that has a large amount of white mixed in it. Hemingway's description of blood in the bullfighting ring indicates a red with a very strong (saturated) chroma.

If you park a red car half in the shade and half in the sunlight, each half would have a different value but retain the same hue and chroma. Only if the hood of the red car reflects a "hot spot" of direct sunlight will that area have a different chroma. Black is a hue with no chroma and a dark value, while white also has no chroma, but a light value.

The concepts of hue, chroma, and value allow language to deal with subtleties in color. They are equivalent to the scale of zones in black and white because they help translate what you see into what the film sees. Thinking in these terms is important because of how your perceptions can change the nature of color that you see.

* Josef Albers' *Interaction of Color* (New Haven, CT: Yale University Press, revised edition, 1975) is an excellent text for the photographer wishing to learn about how color interacts with perception.

† A.H. Munsell, *A Color Notation*, p. 14.

This photograph of the hood of a red car illustrates the different ways of describing color: The car is a single hue, a specific shade of red. The chroma (saturation) of the red color remains constant, except where there is a glare of sunlight. This weakens (desaturates) the chroma of the red by varying amounts, particularly where there is a specular highlight. The lighted portion of the hood shows a light color value, and the shaded portion of the hood (upper left-hand corner) shows a dark color value.

Relativity of Color

In color, as in everything else, perceptions are relative. A classic demonstration of the relativity of perception is to place one hand in a bowl of warm water and the other in a bowl of cold water. After about 20 seconds, move both hands to a third bowl containing lukewarm water. The hand coming from the bowl of warm water will feel cold, while the hand coming from the cold water will feel warm. The way you perceive hue, chroma, and value is also subject to this type of relativity. Simple experiments can help you demonstrate this.

Hue Experiment

Hue cannot exist without a reference. A room illuminated only by blue light gradually appears to be lit by dim, normal color light. A standard trick in the production of night scenes for motion pictures is to shoot in daylight but to filter everything in blue. The eye does not notice the blue color but perceives the scene as dark.

The following experiment will help you understand the relationship of hue to your perceptions:

1 Find an even light source such as a light box or a blank white painted wall.
2 Hold a piece of blue acetate up to your eyes and look at the light source through the acetate until your eyes stop seeing any color.
3 Take piece of a different colored acetate and gradually move it in front of your eyes until you are looking through both the new color and the blue color.

When you are looking at a monochromatic surface through a single color, that color has no reference and you quickly lose the sense of the hue. When you add a point of reference to your field of vision in the form of a second color, the original color reappears along with the new color. Try repeating the experiment using different colors of acetate.

Photographers who make color prints learn to see color imbalances in their prints by glancing at the print quickly and then at a standard reference color. Their eyes adjust to a color imbalance and accept it as correct if they stare at a print too long.

Chroma Experiment

Your perception of color chroma is also relative. For this experiment, you need several pieces of paper, all shades of the same hue.

1. Lay the paper out in strips, from the weakest chroma to the strongest chroma, so that they almost touch.
2. Cover the half of the strips that have the strongest chroma.
3. From the paper strips that are still visible, pick out the one that has the strongest chroma.
4. Cover all but the strip you chose and uncover the rest.
5. Compare the strip you initially chose as having the strongest chroma with the others now visible.

The shade you originally chose as having the strongest chroma now has the weakest. By changing the relationship of color shades, you change your perception of that shade. Repeat this experiment with other colors.

Perception of chroma is relative. Cover the right side of this illustration. Which shade of the remaining colors appears to have the strongest chroma? Next, uncover the right side and cover the left, with the exception of the shade you chose as the one with the strongest chroma. How has your perception of the chroma of that shade changed?

Value Experiment

You can demonstrate how your perception of value changes by performing the following experiment:

1. Take two squares of the same piece of color paper.
2. Place one square in the middle of a larger piece of white paper
3. Place the other square on a larger piece of black paper.
4. Decide which color has the greatest value.

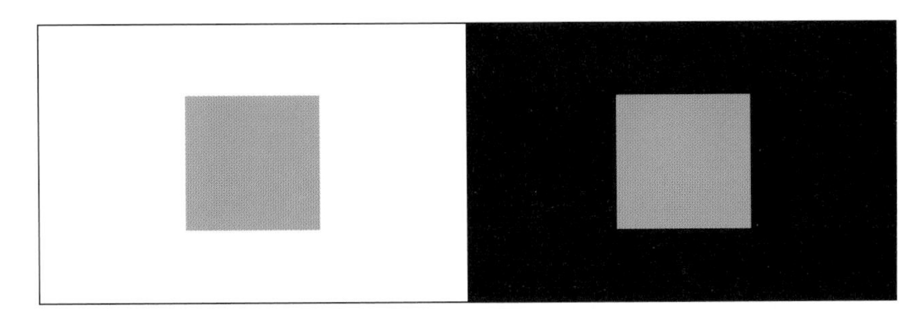

Although the colors in the black squares have the same value (as well as the same hue and chroma), you see them as different because of the difference in the surrounding colors.

The color square on the black field appears to have greater value than the one on the white field. Try this experiment by substituting colors other than black and white for the background colors.

COLOR CONTRAST

Just as you must adapt your vocabulary for describing color, you must also adapt your understanding of contrast as it applies to color. In black-and-white photography, you define contrast as the difference in light reflectance between the highlights and shadows. Differences in light reflectance between colors are often not as important as the contrast between the colors themselves. To understand color contrast, it helps to understand the relationship of the primary colors, both additive and subtractive.

Additive Primaries

White light is made up of three basic or primary colors: red, green, and blue. The best way to imagine the relationship between each color is on an equilateral triangle in which each color takes a point, or in the relationship indicated by the color wheel illustrated below. These colors are called additive primaries because each "adds" its color to the total of white light.

Subtractive Primaries

For every additive primary, there is a complementary color, one that contains all other colors except that additive primary. These complementary colors are cyan (equal parts blue and green), magenta (equal parts red and blue), and yellow (equal parts red and green). Each of these colors subtracts its additive complement from white light and so is called a "subtractive" primary. Cyan is the complement of red, magenta the complement of green, and yellow the complement of blue. Their relationship to the additive primaries are shown in the color wheel.

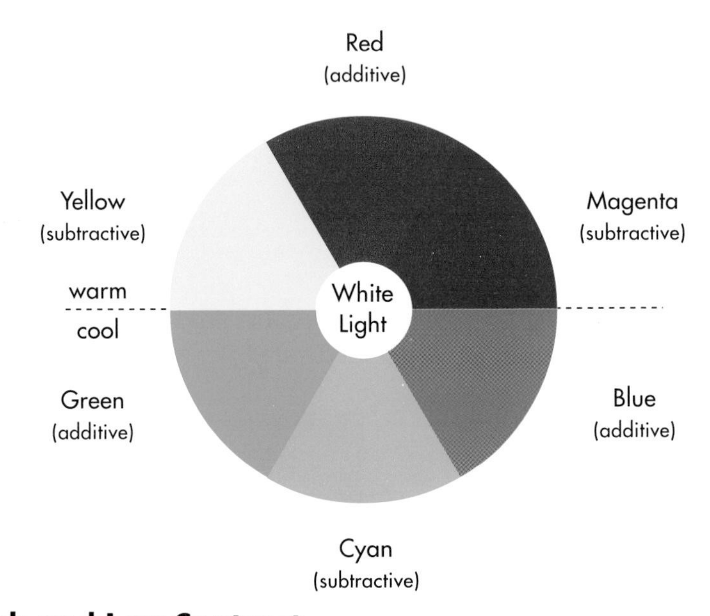

This color wheel shows the relationship of additive and subtractive primary colors. The relationships of these colors determines the contrast of a color image more than differences in light reflectance.

High and Low Contrast

Contrast in color images is higher when colors that are not next to each other on the color wheel appear next to each other in the image. A blue next to a red, or a yellow next to a cyan, for example, gives the appearance of high contrast, even when the two colors reflect the same amount of light.

One reason for high color contrast has to do with the structure of the eye. Each color reflects a different wavelength of light, red being the longest and blue the shortest. The eye must focus differently for each wavelength, and when contrasting colors are next to each other, the eye cannot focus on them both at the same time. The colors seem to vibrate as the eye focuses back and forth.

Color contrast is lower when colors near each other on the color wheel appear next to each other in the image. A green next to a yellow or a blue next to a cyan, for example, gives the appearance of low contrast. This happens even when the two colors reflect significantly different amounts of light. The eye can comfortably focus on both colors at the same time.

Warm And Cool Contrast

When adjacent colors in an image cross the boundary between the warm colors (red, yellow, and magenta) and the cool colors (blue, green, and cyan), it enhances the appearance of contrast. This is mostly a psychological effect. People tend to associate shades of red, yellow, and magenta with warm feelings, such as the heat created by the light of the sun or a fire. Blue, green, and cyan, by contrast, usually evoke a sense of cool water, forests, or even ice.

Color Contrast Experiment

To demonstrate the difference between contrast in color and contrast in black and white, try this experiment:

1. Collect eight or more pieces of paper that each have a single solid color. Construction paper will work well. Use only pieces that are large enough to read with your light meter.
2. Measure the light reflectance from each piece of paper and group the colors by the amount of light they reflect. Within a group, no piece of paper should reflect half a stop of light more or less than any other.
3. In each group of colors, select two that to your eyes have the most contrast and place them side by side.
4. Visualize these colors as they would appear in a black-and-white photograph.

Since the indicated meter readings from each piece of paper are no more than half a stop apart, in a black-and-white print they will appear as virtually the same shade of gray. This would be a low contrast image. Yet these colors are the ones that had the greatest contrast to your eyes.

USING COLOR MATERIALS

The two basic film choices for making color photographs are transparency film and negative film. Each is different, primarily in the way you view the final image. This difference affects the way you previsualize and place the tones in the scene.

You view transparencies by transmitted light, whether projected in a darkened room or held up against a light box. Light passes through the image once on its way to your eye. You view prints by light that reflects off the base support material. Light passes through the image in the emulsion twice, once going and once coming. This doubles the visible density of the print image and limits the range of tones you can see.

You can see a wider range of tones in a transparency than in a print. This is not because of any significant differences in the two emulsions, but rather because of the way you view the images. Light passes through the tones of a transparency only once before you see the image. Light must pass through the tones of a print twice, once before it strikes the print's base material and a second time after it reflects off the base. This doubling of density reduces the number of tones that you can see in a print compared to a transparency.

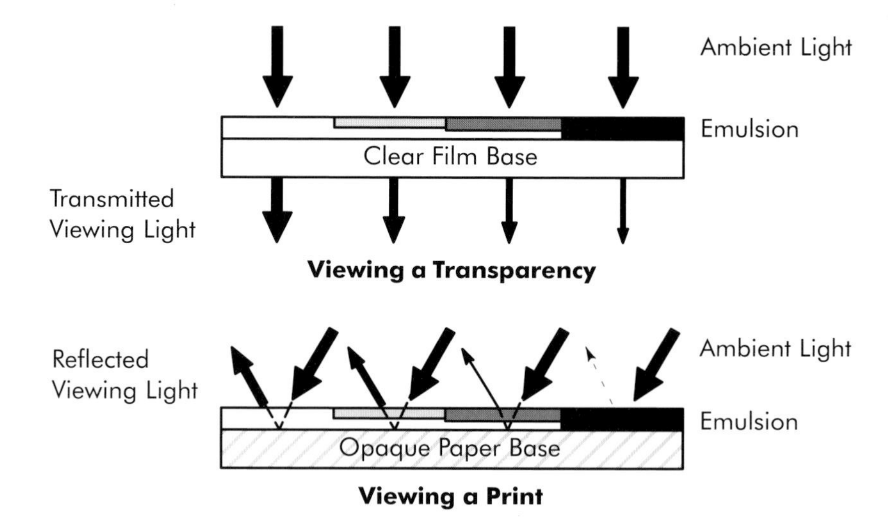

Transparencies

Compared with a print, a transparency offers a much greater range of tones and subtlety. This is fortunate because, unlike a negative, you cannot easily manipulate the tonal range of a transparency through changes in development time. When you alter the processing time of color film, you risk causing changes to the film's color reproduction (known as *color balance*). This makes the color contrast in a scene more important than the overall difference in light reflectance.

Exposing Transparency Film

You must be careful when exposing a transparency because the film in the camera is also the final product. You do not have the secondary step of printing to correct for an underexposure or overexposure.

Correct exposure for transparency film follows the same principle as for negatives. Exposure has the greatest effect on the areas of least density in the image. In a negative, these are the shadows. In a transparency (which is a positive image), these areas are the highlights.

This means that when you photograph with transparency film, you select the most important highlight in a scene to previsualize and place. Up to certain limits, all the tones darker than the highlight you place will appear in the image. This is possible because of the transparency's ability to render a wider range of tones than a print.

When you use the hand-in-light/hand-in-shadow method for determining the exposure for a transparency, take the indicated meter reading from the palm of your hand in the brightest light and place it in zone VI. As long as you are in the same light as your subject, the exposure will be correct. Contrast is primarily a matter of the colors in the scene.

Exposure Index Test for Transparencies

Base your exposure index for transparencies on a correct rendering of a previsualized zone VII. This is the least dense highlight that still shows identifiable detail. Zone VIII has less density, but this density is too close to that of clear film for you to easily distinguish it from a zone IX in the test.

Find the correct exposure index for the transparency film you want to test in the following manner:

1. Locate a standard zone VII value, such as a textured white sweater or a white-painted brick wall. Compose a portrait with your subject next to the zone VII.
2. Set your light meter to an exposure index that is eight times the film's ISO rating. (For example, if you are testing ISO 200 film, start with a test index of 1600.)
3. Take a meter reading of the zone VII object and open up two f-stops (or shutter speed equivalents) to place the indicated meter reading in zone VII. Make the exposure.
4. Repeat step 3, making a series of exposures that tests at least seven different indexes, from eight times more than the film's ISO to eight times less. (For example, if you are testing ISO 200 film, use test indexes of 1600, 800, 400, 200, 100, 50, and 25.)
5. Develop the film according to the manufacturer's recommendations. You can send the film to a commercial processing lab.
6. Once the film is processed, arrange the slides in order from the highest test index to the lowest text index.
7. Project the slides in a situation similar, if not identical, to the one in which you will typically view your slides.
8. Choose the frame that best renders the zone VII value, full of texture but not too dark. This frame indicates the correct exposure index.

The most important thing in choosing an exposure index for transparencies is to consider the conditions under which you plan to view the image. The more intense the viewing light, the more density the highlights must have in order to see detail. This means that for the same film (and camera), you could have one exposure index for images you plan to view only on a light box and another for images that you plan to project in a large auditorium.

Underexposure

Correct Exposure

Overexposure

Base your exposure index test for transparencies on the correct rendering of a textured highlight (see page 109). In this image, shot on Kodachrome ISO 64 film, the white cinder block wall was previsualized as a zone VII. Exposure indexes from E.I. 500 to E.I. 8 were tested. These three illustrations, evaluated by projecting them in a darkened room, show an underexposure where the wall appears too dark, a correct exposure where the wall appears as previsualized, and an overexposure where the wall appears too light and lacking in detail.

Contrast Control With Color Transparences

Although contrast in color film is primarily a product of the colors in the scene, there are reasons why you might need to reduce or increase the light reflectance contrast. For example, if you are planning to make prints from a transparency using the Ilfochrome process, you might want to reduce the light reflectance contrast to match the contrast of the print material. Or, if you encounter a scene in which both the color contrast and the light reflectance contrast are very low and increasing the color contrast is not an option, you might choose to increase the light reflectance contrast. Two processes that give the equivalent of increasing or decreasing the light reflectance contrast of color transparencies are *pre-exposure* and *push processing*.

Pre-Exposure

Pre-exposure, also called *flashing*, is a technique that you can use to lower the contrast of transparency film by giving the film an initial exposure to a small amount of light. This light, when added to the overall exposure of the scene, adds significant density to the shadows without noticeably affecting the highlights. To use this technique, you must have a camera that allows for multiple exposures of a single frame.

Use the following procedure as a starting point for pre-exposing your film:

1 Find a scene with a high light reflectance contrast.
2 Place a small piece of $\frac{1}{8}$-inch-thick white, translucent Plexiglass over the camera lens. This diffuses the image in front of the lens and allows you to expose the film to an even tone without detail.
3 Take a meter reading through the Plexiglass.
4 Place the indicated meter reading for the pre-exposure, using the following table:

Contrast of Scene	Zone for Pre-exposure
N−1	Zone I
N−2	Zone II

5 Make the pre-exposure, and then set your camera to produce a second exposure of that frame.
6 Compose the scene in your viewfinder and expose the scene as you would normally, by placing the most important highlight.

With this procedure, you control the density of the highlight through placement and support the shadow density through pre-exposure. One advantage of this technique is that you can pre-expose a roll of film on a frame-by-frame basis. The frames that you do not pre-expose are not affected.

This technique works best when the shadows contain a lot of detail. If the shadow areas in the image have a smooth surface, there is a risk that the shadow will simply look gray and "foggy." For this same reason, you should not attempt to pre-expose for more than a zone II value. Zone III should contain detail, and a pre-exposure for zone III will not have any detail.

Also, the color temperature of light you use for the pre-exposure must be the same as the color temperature of the light in the scene. The easiest way to ensure this is to make the pre-exposure by pointing the camera at the scene you want to photograph. As long as there are no bright light sources in the scene, the translucent Plexiglass will create an even pre-exposure on the film.

Pushing

The common description of pushing color transparency film is to increase the speed rating of the film by increasing the time (or temperature) of the first developer. What actually happens is the same as what happens when you increase the development time of black-and-white film. The highlights increase in density while the shadows remain the same.

This gives you an effective contrast increase of one zone for every "stop" you push the film speed. The drawbacks of doing this include added grain, a slight color imbalance, and of course, no actual increase in film speed.

You can "push" most transparency films that you can process in Kodak's E6 process, such as Kodak's Ektachrome and Agfa's Agfachrome films. In limited situations, you can also push Kodak's Kodachrome film.

To use push processing as the equivalent of high contrast development, previsualize and place the most important shadows in the scene instead of the highlights. (For the hand exposure method, place the hand-in-shadow in zone IV.) The following table describes the combination of contrast, amount of "push," and additional time to add to the first developer for the E6 process:

Contrast of Scene	Stops to "Push"	Add to First Development*
N+1/2	1/2 stop	1 minute
N+1	1 stop	2 minutes
N+2	2 stops	5 minutes

Color Negatives

All color negative films in common use today are compatible with Kodak's C-41 color negative developing process. This is convenient and necessary, because it is difficult to attain the correct balance of

* Marc Levey, *The 35mm Film Source Book*, p. 194.

color dyes in a negative. Industry standardization has been the answer to this problem.

Even if all color negative film is processed the same way, no two brands of film have the same response to color. Each emulsion has a different character, centering primarily around the saturation of certain colors. The best plan is to try the different brands available until you find one that suits your personal sense of color. Do not be awed by the word *professional* on the film label. In some cases, the so-called amateur films produce better results and require less care in handling.

Exposure Index Test for Color Negatives

Expose color negatives, like black-and-white negatives, for the proper shadow density. An exposure index test for color negative film should determine the minimum exposure the film needs to produce adequate detail in a previsualized zone III.

Use the following test to find the best exposure index for color negative film:

1. Find a location in even sunlight and place a gray card and a gray scale in it. The *MacBeth ColorChecker* or the gray card and color/density scale page of the *Kodak Color Dataguide* are both useful for this test.
2. Start with an exposure index that is eight times greater than the film's ISO. (For example, if you are testing ISO 200 film, start with a test index of 1600.)
3. Take your meter readings from the gray card. You do not have to place this reading.
4. Make a series of exposures that tests at least seven different indexes, from eight times greater than the film's ISO to eight times less. (For example, if you are testing ISO 200 film, use test indexes of 1600, 800, 400, 200, 100, 50, and 25.)
5. Process the film according to the manufacturer's recommendations. You can send the film to a commercial processing lab.
6. Identify the index of each frame and place them all in 35mm slide mounts.
7. Project the mounted negatives in a sequence that begins with the highest index, or use a magnifying glass to examine the negatives on a light box.
8. Observe the image of the black-and-white density scale that is part of the page you photographed. The correct index is indicated by visible separation in all the steps.

Each step in the density scale is a separate reflection density. As you view the negatives starting from the highest index, you will see more and more steps in the density scale becoming separate tones. The point at which each step is a separate density indicates the correct exposure index.

If you have difficulty deciding between two adjacent frames in the test, always decide in favor of the frame with the most exposure. Unlike the problems that excessive density causes in black-and-white negatives, color negatives are not adversely affected if there is only moderate overexposure.

Underexposure

Correct exposure

Overexposure

An exposure index test for color negative film (described on page 112) is based on an examination of shadow values. Underexposure causes the density scale (the narrow vertical column) to have no separation in the bottom segments. Correct exposure causes all segments of the density scale to be separate, distinct tones. Overexposure adds unnecessary density throughout the negative.

Contrast Control With Color Negatives

Although it usually is not necessary, you can control the contrast between highlights and shadows in color negatives by modifying the Kodak C-41 process. Kodak does not recommend it, but you can make small changes in the film's development time with a minimum effect on the balance of the color dyes. The result is the ability to change the contrast by a one-zone expansion or contraction.

The following times are based on Kodak's recommended normal development time for the first roll of film processed in fresh C-41 developer:*

Contrast of Scene	Development Time
N−1	2 minutes, 40 seconds
N	3 minutes, 15 seconds
N+1	4 minutes

This image, shot on transparency film, illustrates color contrast among the primaries of cyan, red, green, and yellow. To make this photograph, the photographer metered her hand and placed it in zone VI. Since the highlights in the scene were too distant and the moment too brief to walk up and meter directly, this was the only practical method.

* Arnold Gassan, *The Color Print Book* (Rochester, NY: Light Impressions, 1981, p. 67).

SUMMARY

Our perception of color is affected by the relationship of each color to another and is one reason why our perception of color is relative. Any understanding of the zone system and color photography must start with an understanding of the interaction of color.

- Every color has three significant features that define it: *hue, chroma,* and *value*:
  - *Hue* is the name of a color, such as green or red. It refers to a specific wavelength or part of the color spectrum.
  - *Chroma* indicates the purity of a hue. A pastel color has a weak chroma. Vivid hues have a strong chroma.
  - *Value* refers to the intensity of a color as it appears next to other colors in a scene.
- The concepts of hue, chroma, and value are equivalent to the scale of zones in black-and-white photography because they help translate what you see into what color film sees.
- Color contrast is defined by how primary colors relate to each other on the color wheel. Differences in light reflectance between colors are often not as important as the contrast between the colors.
- Expose transparency film by previsualizing and placing the most important highlight. Determine an exposure index by finding the exposure that best renders a zone VII highlight the way you normally view transparencies.
- To decrease the light reflectance contrast of a color transparency, pre-expose the film.
- To increase the light reflectance contrast of some color transparency films, "push" process the film.
- Expose color negative film for the most important previsualized shadow. Determine an exposure index by shooting a scale of gray tones to discover the exposure that best renders the darkest tones in the gray scale as individual steps.
- With the C-41 color negative process, you can vary the development time within limited parameters to achieve as much as an N−1 to N+1 contrast change.

Afterword

In the final analysis, the only way to master the zone system is through practice. The techniques of previsualization, metering, and contrast determination do not simply yield themselves on the first attempt. The zone system is most valuable as a tool when you learn it thoroughly enough to use it spontaneously. Minor White, a pioneer of teaching the zone system, even suggested that you have only begun to master it when you start asking people in what zone they would like their morning toast.*

Much has been made in this volume about learning the terminology of the zone system as one would learn another language. This involves not simply rote definitions but the ability to think in the new language. When you make the effort to learn, a totally new form of expression opens up to you. No longer must you default your creativity to the camera machinery and the chemical process. Instead, you can assert increasing control over your materials.

This freedom of choice that the zone system offers also implies responsibility. Previsualization ultimately involves thinking not just about print tones, but also about all other aspects of your photographs, including how they will affect the viewer. This holistic view of photography is what Minor White called a higher form of doing.† Once you master the zone system, you may find that you are a different photographer than you were when you started.

---

* Minor White, *The Zone System Manual* (Dobbs Ferry, NY: Morgan & Morgan, 1968, p. 59).

† Minor White, *op. cit.*, p. 98.

Appendix A

Glossary

18% Reflectance The percentage of light reflected from a zone V medium-gray tone, such as the one represented on the Kodak Neutral Test Card.

Acutance The degree of a negative's sharpness. High acutance film developers enhance the appearance of sharpness in a negative.

Additive Primaries The three colors that make up white light: red, green, and blue. Each primary adds its color to the total of white light. See also *Subtractive Primaries*.

Alternative Placement Using previsualization to choose different exposures based on shadow placement (for negatives) or highlight placement (for transparencies). Alternative placement is a way to explore different ways of rendering the placed zones in a scene. See also *Bracketing*.

Aperture The size of the hole created by the adjustable diaphragm that controls the light-transmitting ability of the camera lens.

Average Meter Reading The default reading given by a reflected light meter, one that averages the light reflectance from all the highlights and shadows in a scene. An average meter reading does not give any indication of contrast.

Bracketing Photographing a scene using three or more different exposures (usually one stop more and one stop less than an average meter reading) in hopes of getting at least one printable negative.

Brightness See *Value*.

Callier Effect Light scattered by a silver emulsion in proportion to its density when you enlarge a negative with a condenser enlarger. The Callier effect adds contrast to an image.

Chroma (also known as *saturation*) indicates the strength or intensity of a hue. A hue mixed with any other hue, or black or white, will change the chroma, or intensity of that hue. A pastel color, which has a hue mixed with a lot of white, has a weak chroma. Vivid hues have a strong chroma.

Cold-Light Enlarger A type of diffusion enlarger that uses a specially designed fluorescent tube as a light source. See also *Diffusion Enlarger*.

Color Balance An indication of accurate reproduction in color film. Each of the three color layers, cyan, magenta, and yellow, must be balanced to properly reproduce a scale of gray tones.

Condenser Enlarger A type of enlarger that uses large glass lenses above the negative to concentrate and focus the light before it passes through the negative. See also *Diffusion Enlarger, Cold-Light Enlarger,* and *Callier Effect.*

Contrast The difference between the light reflectance of highlights and shadows in a scene. In black-and-white images, contrast is the difference between the highest and lowest densities. In color images, contrast is primarily a function of the arrangement of hues. See also *Warm and Cold Contrast.*

Diapositive See *Transparency.*

Density The ability of a film negative or a print image to absorb light. The more reduced silver in a given area, the greater the light absorption and the greater the density. See also *Film Base Density.*

Densitometer A device for measuring density. The measurement is usually in the form of a logarithmic scale. A densitometer for measuring film densities is a *transmission densitometer.* A densitometer for measuring print densities is a *reflection densitometer.*

Depth of Field The distance between the nearest and farthest objects in the scene that are in acceptable focus. The smaller the lens aperture, the greater the depth of field; the larger the lens aperture, the smaller the depth of field. See also *Aperture.*

Development The chemical reaction that reduces the electrical state of exposed silver halides to visible, metallic silver particles.

Diffusion Enlarger An enlarger type that uses a piece of frosted glass or plastic just above the negative to scatter the light nondirectionally before it passes through the negative. See also *Condenser Enlarger* and *Cold-Light Enlarger.*

Exposure Index The calibration you use on your light meter to compensate for a consistent error in your light meter or your camera shutter.

Exposure Placement See *Placement.*

Film Base Density The least density possible on developed film. The film base density includes the density of the film support material plus any developed residual silver, known as *fog.*

Flashing See *Pre-exposure.*

Fog The residual density in unexposed, but developed film.

Gray Card A piece of cardboard or other stiff material with a gray coating that reflects exactly 18% of the light falling on it, such as the Kodak Neutral Test Card.

Silver Halide The light sensitive gains of silver on a piece of film or printing paper, which, when exposed to light, contain a latent image. The visible image in film consists of exposed silver halides reduced to metallic silver during development.

High Contrast A scene in which the difference in indicated meter readings is greater than the difference in previsualized zones for the highlights and shadows in that scene.

Highlight Light tones in a scene or a print, or dark densities in a negative. In the zone system, any zone lighter than zone V is a highlight.

Hue The name of a color, such as green or red which represents a specific wavelength or part of the color spectrum.

Hyperfocal Distance Setting the focus for the closest distance at which the depth of field also includes the infinity mark on the lens barrel. On a bright day, hyperfocal distance might include everything from 10 feet to infinity.

Incident Light Meter A type of light meter that reads the light falling on a scene rather than the light reflected from the scene.

Indicated Meter Reading A measurement of reflected light from a previsualized tone. An indicated meter reading renders any previsualized tone as zone V. See also *Average Meter Reading*.

Latent Image The potential for a visible image on film after undergoing the chemical process known as development.

Law of Reciprocity The one-to-one (reciprocal) relationship between the lens aperture and shutter speed. A change in either the aperture or shutter speed changes the exposure by a factor of two. You can change the f-stop and keep the same exposure by making a corresponding opposite change in the shutter speed. See also *Reciprocity Failure*.

Light Reflectance The amount of light from a light source that is redirected by the surface of an object. Light reflectance is usually measured with a light meter calibrated to a standard value of 18% reflectance.

Low Contrast A scene in which the difference in indicated meter readings is less than the difference in previsualized zones for the highlights and shadows in that scene.

Luminance See *Value*.

Medium Gray Any object reflecting approximately 18% of the light that falls on it. In the zone system, medium gray is zone V. See also *Gray Card*.

Meter Reading See *Indicated Meter Reading*.

Midtone See *Medium Gray*.

Normal Contrast A scene in which the difference in indicated meter readings matches the difference in previsualized zones for the highlights and shadows in that scene.

Placement The process of using your camera's controls to record your shadow previsualization on negative film, and ultimately on the print. You place highlights when using transparency film.

Pre-exposure (also known as *flashing*) A technique to lower contrast by giving a light-sensitive emulsion (such as film) an initial exposure to a small amount of light. This exposure adds a significant density to the shadows without noticeably affecting the highlights. Useful with transparency film where you cannot easily change contrast through development.

Previsualization Establishing a mental image of the tones on a print before you expose the film.

Proper Proof A contact sheet you create in a specific way to yield consistent information about film exposure and contrast.

Pushing A method for increasing the contrast of transparency film by increasing the time (or temperature) of the first developer. Does not increase the light sensitivity of film as is commonly thought.

Reciprocity See *Law of Reciprocity.*

Reciprocity Failure The failure of the law of reciprocity which occurs when an exposure is either extremely long or short, or when the light intensity is extremely high or low. More exposure is necessary in these situations than the law of reciprocity predicts. See also *Law of Reciprocity.*

Reflectance See *Light Reflectance*.

Reflected Light Meter Type of light meter that reads the light reflected from a scene.

Reflection Densitometer See *Densitometer.*

Representational Print A print in which all the tones look the way the eye sees them in the scene without alteration.

Saturation The term color photographers usually use for the intensity of a hue. See also *Chroma*.

Shadow Dark tones in a scene or print, and light densities in a negative. In the zone system, any zone darker than zone V is a shadow.

Slide See *Transparency.*

Subtractive Primaries The colors that subtract their complementary additive primary color from white light. Cyan contains every color except red, magenta contains every color except green, and yellow contains every color except blue. See also *Additive Primaries*.

Transmission Densitometer See *Densitometer.*

Transparency A positive image on film viewed by transmitted light, commonly called a *slide* and sometimes called a *diapositive*. Can be either a black-and-white or color image, though usually thought of as a color image.

Tones How film densities appear on a print. The lighter the tone in a scene, the greater the film density and the lighter the tone on the print.

Value The lightness or darkness of a color as it appears next to other colors in a scene.

Warm and Cool Contrast When adjacent colors in an image cross the boundary between the warm colors (red, yellow, and magenta) and cool colors (blue, green, and cyan), the appearance of contrast is enhanced. This is more of a psychological effect rather than a visual effect.

Zone Focus Using an estimate of the distance to the main subject for setting the lens focus and allowing the depth of field to make up for slight inaccuracies.

Zones In the zone system, the division of the continuous tone gray scale into nine distinct tones ranging from black to white. Each zone is related to the zones adjacent to it by an exposure factor of one f-stop or one shutter speed. Zones are the units that photographers use to translate what their eyes see into photographic procedures.

Appendix B

Shutter Consistency Test

If you suspect that your camera shutter is producing inconsistent exposures, as described on page 70 of Chapter 8, *Choosing Your Exposure Index*, you can perform the following test to confirm your suspicion. For this test, you need a gray card (Kodak Neutral Test Card or equivalent) and a tripod in addition to your camera and a roll of film.

1. Place the gray card in even light and position your camera on the tripod in front of it. Move the camera close enough to the gray card so that the out of focus gray image fills the viewfinder without casting a shadow on the card.
2. Take a meter reading from the gray card and place the indicated reading in zone II.
3. Shoot three frames at the same shutter speed and f-stop combination.
4. Change the shutter speed and compensate by changing the f-stop according to the law of reciprocity to maintain a zone II placement. Expose three more frames.
5. Continue until you have shot three frames at every possible shutter speed with the available f-stops on that lens.
6. Repeat steps 3 through 5 until you finish the roll of film.
7. Develop the film and make a proper proof of the negatives. (See Chapter 8 for a description of how to make a proper proof.)

Inspect the images of the gray card on the contact sheet. Look for a consistent, dark-gray tone on all the frames. A small amount of density variation between frames is acceptable, but anything beyond that is a cause for concern. If this test indicates that your comera is making inconsistent exposures, have the shutter checked and repaired if necessary.

Appendix C

DX Coding of Film Cassettes

Not that long ago, the information on the exterior of 35mm film cassettes was intended for the eyes only. Modern film cassettes contain information not only for your eyes, but also for automated cameras and film processing labs. Every cassette from a major manufacturer contains machine-readable codes that describe the specific film emulsion, the ISO film speed rating, number of exposures, and the exposure range tolerance of the film inside.

The following diagram illustrates this coding:

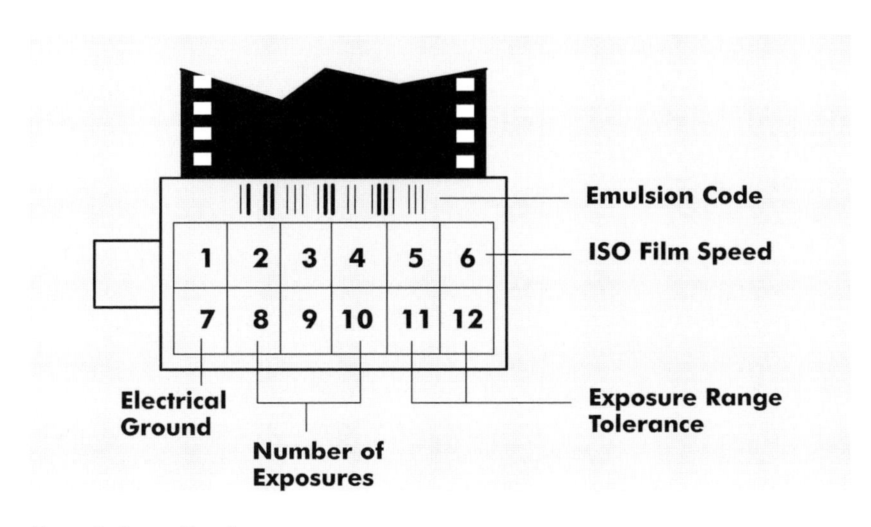

Current 35mm film cassettes contain coding to aid automated film exposure and processing. Knowing these codes enables you to override them when necessary to obtain a correct exposure.

Emulsion Code

The UPC (Universal Product Code) bars on the cassette describe the specific film emulsion in the cassette. Many automated processing machines can read this code and make adjustments accordingly, either during film development or printing. Some film emulsions also have this code embedded as a latent image on the film edges.

DX CODES

The pattern of 12 silver and black rectangles on the outside of the 35mm film cassette are the DX codes. Electrical contacts inside automated cameras use these codes to determine the ISO speed rating of the film. Some cameras also use the DX codes to determine the film length and the ability of the particular emulsion to tolerate exposure errors.

You can tell how much of the DX coding your camera utilizes by counting the number of electrical contacts on the inside where you place the film cassette. These contacts are usually small, rounded silver bumps that make contact with the cassette at one of the code positions.

- If there are four contacts, the camera can read ISO film speeds in full f-stop increments only.
- If there are six contacts, the camera can read the full range of ISO film speeds in increments of roughly one-third of an f-stop.
- If there are 12 contacts, then the camera can take advantage of the full range of DX coding information, including film length and exposure tolerance.

The following sections describe the DX codes in detail.

Electrical Ground

Positions 1 and 7 are always silver (open to electrical conductance) to provide a ground for the contacts.

ISO Film Speed Ratings

Positions 2 though 6 describe the light sensitivity of the emulsion using the ISO rating standard. The following chart shows the DX code combinations from ISO 25 to ISO 5000.

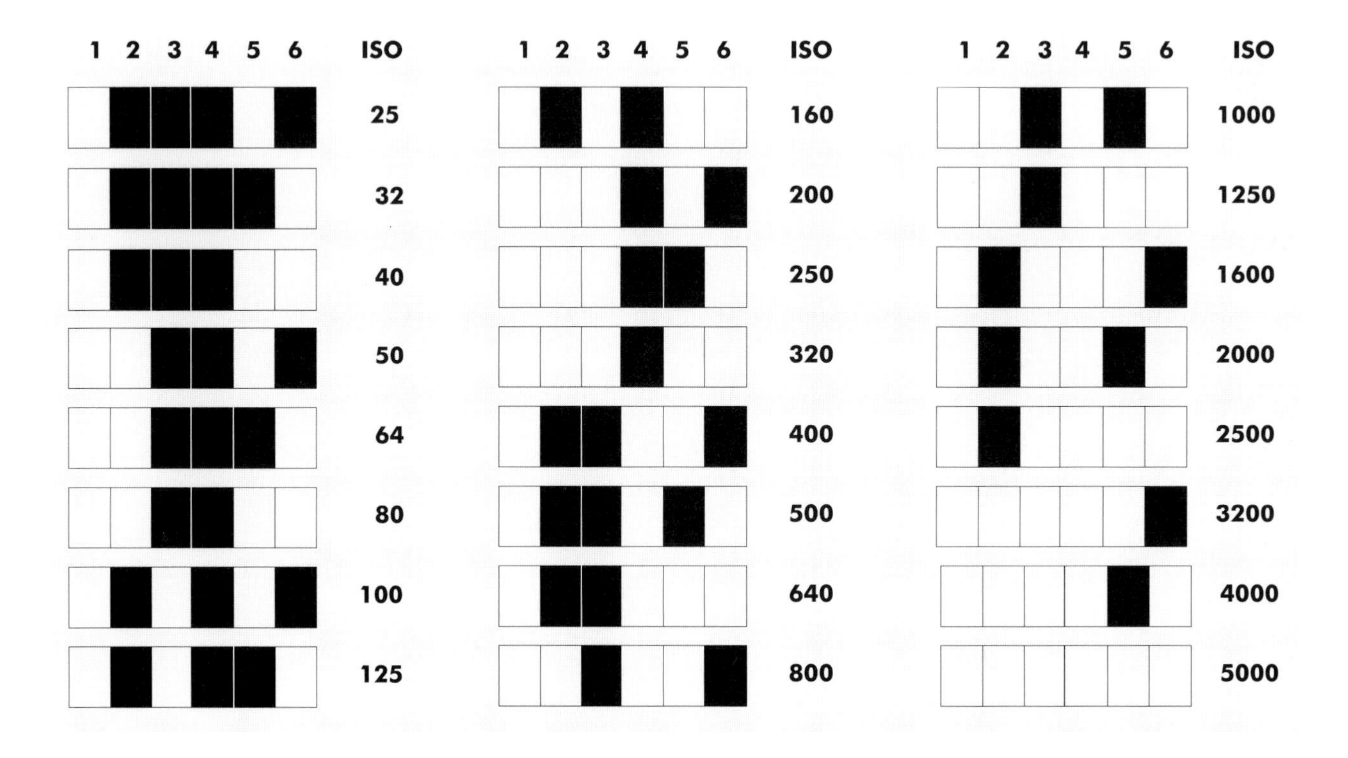

Film Length

Positions 8 through 10 indicate the film's length. For cameras properly equipped, this allows the camera to signal the end of the roll. The following chart illustrates these codes:

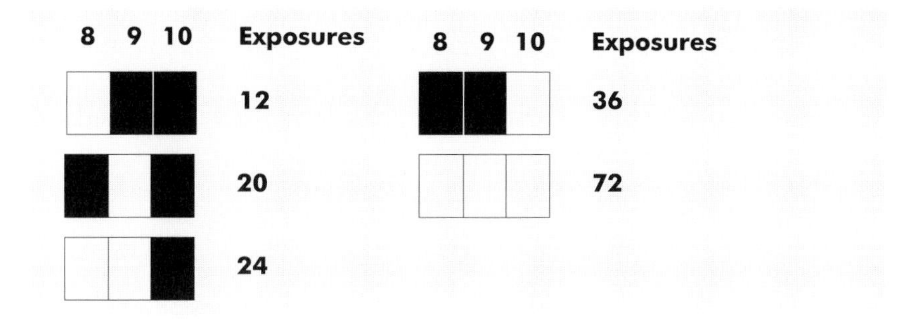

Film Latitude

Positions 11 and 12 indicate the film's latent image tolerance for exposure. This helps properly equipped cameras determine the difference between transparency, color negative, and black-and-white negative films. The following chart illustrates these codes:

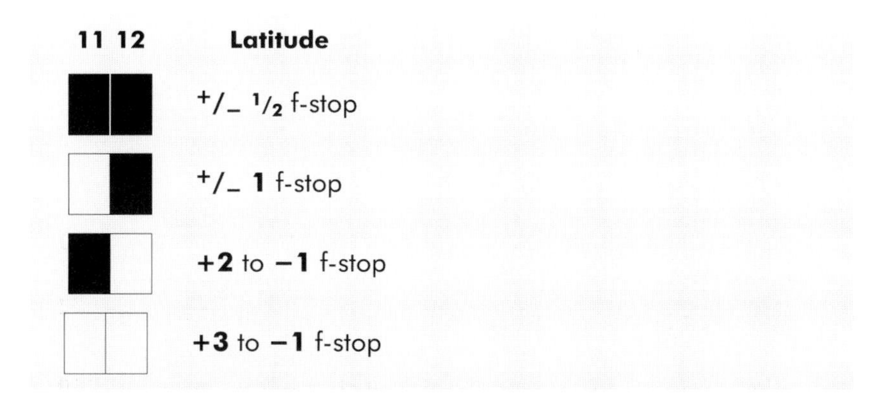

OVERRIDING THE DX CODES

If your exposure index for a film is different from its ISO film speed rating, you can alter the DX codes on the film cassette to make that exposure index work in an automatic camera that has no manual override. The number of contacts in your camera determines how much information the camera can use. If your camera has only four contacts (in positions 1 through 4), you can change the exposure index by whole stop increments only.

You can purchase gummed labels with new ISO settings to paste over positions 1 through 6. One source for DX-coded film labels is Porter's Camera Store, Inc., Box 628, Cedar Falls, IA 50613-0628. You can also create your own labels by adhering conductive silver mylar tape over positions 1 through 6 and then adding nonconductive black electrical tape over the positions blacked out in the diagram on page 122. Appropriate silver tape is 3M slide masking tape, catalog number 91-30610, available from most camera stores.

Appendix D

ASA, DIN, and ISO Film Speed Ratings

There are three different scales that measure the light sensitivity of film. The most commonly used scale is the one established by the International Standards Association (ISO). This arithmetic scale indicates a doubling of light sensitivity when the rating number doubles. For example, ISO 200 film is twice as sensitive to light as ISO 100 film. The ISO scale is identical to the now obsolete American Standards Association (ASA) system. You may still find some light meters and films calibrated to the ASA scale.

The Deutches Industrie Norm (DIN) has established a third rating scale. This is a logarithmically derived scale that indicates a doubling of light sensitivity when the scale increases three numbers. For example, DIN 24 film is twice as sensitive as DIN 21 film.

Use the following chart to compare the three systems:

ASA	DIN	ISO	ASA	DIN	ISO
10	10	10	320	26	320
12	12	12	400	27	400
16	13	16	500	28	500
25	15	25	640	29	640
32	16	32	800	30	800
40	17	40	1000	31	1000
50	18	50	1250	32	1250
64	19	64	1600	33	1600
80	20	80	2000	34	2000
100	21	100	2500	35	2500
125	22	125	3200	36	3200
160	23	160	4000	37	4000
200	24	200	5000	38	5000
250	25	250	6400	39	6400

Index

D

E

F

G

H

R

S

T

Z

Photo Credits

All photographs by the author unless credited.

Cover	Judith Canty
Page 1	Judith Canty
Page 3	Judith Canty
Page 5	Judith Canty
Page 11	Judith Canty
Page 12	Bill Byers
Page 13	Lois Lord
Page 14	David Torcoletti
Page 16	Lois Lord
Page 17	David Torcoletti
Page 18	Erika Sidor
Page 20	Lois Lord
Page 21	Bill Byers
Page 26	Judith Canty
Page 35	Judith Canty
Page 37	Judith Canty
Page 38	Judith Canty
Page 41	Judith Canty
Page 48	Erika Sidor
Page 59	Tim Barnwell
Page 85	Judith Canty
Page 86	Judith Canty
Page 91	Judith Canty
Page 92	Bill Byers
Page 94	Judith Canty
Page 101	Cheryl R. Sacks
Page 113	Judith Canty